Angelica Buzinaro Avaci

Economic evaluation of biogas production with swine waste

Angelica Buzinaro Avaci

Economic evaluation of biogas production with swine waste

Financial assessment of biogas production from pig farming

ScienciaScripts

Imprint

Cover image: www.ingimage.com

This book is a translation from the original published under ISBN 978-613-9-68668-1.

Publisher:
Sciencia Scripts
is a trademark of
Dodo Books Indian Ocean Ltd. and OmniScriptum S.R.L publishing group

120 High Road, East Finchley, London, N2 9ED, United Kingdom
Str. Armeneasca 28/1, office 1, Chisinau MD-2012, Republic of Moldova, Europe
Printed at: see last page
ISBN: 978-620-8-22242-0

Summary

ACKNOWLEDGMENTS

To GOD for his constant presence in my life and for all the blessings he has given me, always guiding me along the best paths and with the best people.

To my family, especially my parents, who have always supported me in the decisions I have made.

To my husband Fernando, for his support in the happy and sad moments of my life.

To Prof. Dr. Samuel for his patience and guidance throughout my master's degree.

The secretaries of the master's program, Vanderléia Stockmann Schimidt and Tatiane Alves Pidorodeski, for their friendship, encouragement and attention.

To all the professors of the Master's Degree Program in Energy in Agriculture, for the teachings transmitted.

The Coordination for the Improvement of Higher Education Personnel (CAPES), for the financial support during the period of the Master's Degree in Energy in Agriculture.

To all my friends who, in one way or another, encouraged me to finish this stage of my life.

Thank you so much!

GENERAL INTRODUCTION

Biomass has the potential to become one of the most important renewable energy alternatives in the current scenario. It has an important role to play in managing and mastering climate policy today. Integrated bioenergy systems can contribute to securing future energy supplies and provide a diversified energy matrix for both developed and developing countries.

To mitigate climate change, bioenergy acts by reducing carbon dioxide and by progressively replacing fossil fuels with sustainable energy in the future. According to the Kyoto Protocol, the European Union has a target of reducing carbon dioxide emissions by 8% of the 1990 level between 2008 and 2012. It is also committed to increasing the share of renewable energies to 12% of gross energy consumption (Malkki and Virtanen, 2003; COM, 1997).

In Brazil, based on the Kyoto Protocol, federal and state programs have been set up to encourage sustainable food production, reducing the environmental damage caused by such activities.

Livestock production, especially of pigs, generates a large amount of waste which, when not treated properly, causes serious problems for the environment, such as pollution of rivers and springs, soil pollution and unpleasant smells in the vicinity of farms.

Federal incentives and increasingly strict environmental laws have led pig producers to invest heavily in waste treatment in order to reduce the environmental impact generated. The main piece of equipment is the biodigester. In the absence of oxygen, the bacteria degrade, producing methane and carbon dioxide in large quantities, giving rise to biogas.

Biogas is used on farms to meet the energy needs of various segments of the property. One of these is electricity. When a motor generator is installed, it transforms the biogas into electricity.

The Granja Colombari unit, located in the municipality of Sao Miguel do Iguaçu - PR, has a pioneering biogas energy generation project. The property raises pigs in the finishing stages. They arrive at the property weighing 20 kilos and leave weighing approximately 120 kilos.

The aim of this work is to carry out a financial analysis and energy balance of electricity production, which is essential for analyzing the viability of the project. It is divided into three chapters.

Chapter I - Biomass, biogas and bioenergy: describes the bioenergy scenario in Brazil, highlighting the renewal of the national energy matrix.

Chapter II - Financial analysis of microgeneration of electricity via pig biogas: this chapter analyzed the financial aspects of energy production. The number of pigs, manure production, initial investment and biogas production were taken into account, ending with the cost per MWh of energy produced.

Chapter III - Energy balance of biogas production in a pig finishing system: the energy balance of the defined pig production system was carried out. It considers input items (feed, energy, human labor, buildings and water consumption) and output items (live weight of piglets, biogas and biofertilizer).

Chapter I - Biomass, Biogas, Bioenergy

Summary

Bioenergy production is gaining momentum in Brazil. The need to renew the national energy matrix has led to the emergence of various forms of energy production from numerous sources, such as solar, wind, biomass, geothermal and others. The country has a tradition of producing clean energy, but this is not enough to meet the needs of industries, businesses and homes. This chapter aims to show the panorama of bioenergy production in the country. Its characteristics, its forms of production and the incentives provided by the federal government. Biogas comes into this context because it is a renewable and practically infinite source. With the demand for food increasing and the need to produce meat products, there are environmental problems due to the waste generated. Treating waste with biodigesters solves the environmental problem by preventing soil and river contamination and also produces biogas which is used to produce electricity and for use in rural homes. It also produces biofertilizer, which is used on pastures, reducing production costs.

1. Introduction

The constant blackouts in various regions of the country in recent years have raised the question of Brazil's self-sufficiency in electricity generation. Hydroelectric power generation accounts for approximately 74% of total production. However, constant droughts in the country's major river basins have led to a decrease in energy production, generating an energy deficit for certain regions of the country.

The need to expand and renew Brazil's energy matrix has led to other sources of clean, renewable energy gaining strength on the national scene. This is the case with biogas. There are various ways of obtaining biogas, including urban waste and rural waste. In the field of rural waste, the production of pig and cattle waste stands out, which used to be a huge nuisance for producers and can now be used to generate electricity as long as the waste is treated correctly.

This chapter aims to give an overview of bioenergy production in the country, its characteristics and challenges, with an emphasis on energy production with biogas.

2. Bibliographical review

2.1 Alternative energy sources

In order to reduce human dependence on fossil fuels and consequently reduce climate change, we need to switch to a completely renewable energy system that does not emit greenhouse gases and does not harm the environment (Cornelissen et al. 2012). Biomass energy is playing a leading role in replacing fossil fuels with renewable sources (Karpenstein-Machan, 2001).

The growing demand for energy worldwide has led nations to encourage an increase in alternative sources. Brazil is a pioneer in clean energy. The country's main energy matrix is hydroelectricity, which is a renewable and sustainable source. However, the problem with hydroelectric generation is the amount of farmland it occupies, due to the damming up of water. With a view to finding new sources of alternative energy to meet the country's needs in the coming years, the government has been encouraging public and private initiatives to find new ways of producing energy.

Millions of tons of solid waste are produced every year from industrial, urban and agricultural sources. The decomposition of this organic waste has resulted in large-scale contamination of land, water sources and the air. Of all the forms of solid organic waste, the most abundant is animal manure, mainly from small farms, and it is from these farms that the problem of pollution from waste disposal is most intense. Continued research into the treatment of animal manure for the production of biogas and the optimization of methods can be used to increase production and the wider application of the technology (Nasir et al. 2012).

A little over 100 years ago, biomass declined in its leadership as an energy producer, being replaced by coal, oil and natural gas. The use of biomass was reduced to small farms. However, with the need to reduce consumption of fossil fuels, residual biomass has become the most important source of renewable energy (Cortez et al. 2008). According to Costa (2006) the diversification of the energy matrix is imminent and biomass is largely responsible for this change.

According to Marques (2012), alternative energy sources, especially biomass, offer solutions for increasing environmental sustainability. They also solve part of the problems linked to the use of animal and organic waste in urban areas, with the production of electricity using the biogas generated by their anaerobic digestion.

Sources of residual biomass include non-ligneous vegetable waste and woody vegetable waste such as wood and its residues, as well as organic waste, which includes agricultural, urban and industrial waste. However, when looking for the specific availability of energy biomass in a given region, it is important to consider ecological, economic and technological issues (Fernandes, 2012).

According to Goldemberg and Lucon (2008), biomass has great potential to contribute to the energy supply over the next few years. Forecasts show that biomass will rise from a level of 1,313Mtoe in 2010 to 3,271Mtoe in 2040, an increase of more than 100% over 30 years. The International Energy Agency (IEA) estimates that in the next 20 years, approximately 30% of the total energy consumed by humanity will come from renewable sources. Renewable energy currently accounts for 14% of the world's energy supply.

2.2 **Biogas production**

Biogas is an emerging renewable energy source derived from the conversion of plant biomass and organic waste into biofuels through anaerobic decomposition. Over the last decade, the number of biogas plants has increased significantly in industrialized regions (Ward et al. 2008). The development of biogas production has resulted in an increase in the production of waste that is applied as biofertilizer in agriculture (Arthurson, 2009).

The source of biogas generation is divided into two forms: anaerobic digestion of organic waste and gasification. In anaerobic digestion, biogas is produced when microorganisms degrade organic materials in the absence of oxygen. The raw material can be any organic fraction of domestic waste or agricultural waste (pigs, cattle, poultry litter, vinasse, among others) (Lantz et al. 2007). In gasification, woody and non-woody vegetable waste can be used and burned in gasifiers to produce biogas. In addition to reducing greenhouse gases (GHG), the biogas generation system, among other things, reduces eutrophication, air pollution and improves the use of agricultural nutrients.

Compared to other bioenergy systems, biogas production systems are considered to be the most complex, involving several players, such as municipalities, farmers and energy companies, along with factors that influence the system, acting as forms of incentives or barriers. However, the socio-economic benefits and economic viability of biogas production are limited under the political aspects of the country (Lantz, 2004; Svensson et al. 2005). Incentives for biogas production have progressed over the years. Several federal programs have been instituted to encourage sustainability in agricultural production. One of the objectives of the Low Carbon Agriculture Program (ABC Program) implemented by the Federal Government is to encourage the treatment of animal waste, reducing environmental impacts and reducing GHG emissions.

From an environmental point of view, biogas production systems are complex to study, because the environmental impacts are unique to each system. Emissions from biogas production vary according to the raw material and the composition of the gas that is produced. When assessing the environmental impact and impacts of using biogas, it is necessary to include end-use emissions and emissions from the fuel production chain. The variations make it impossible to draw any conclusions about the environmental impact of biogas production (Borjesson and Berglund, 2006).

Biogas differs from traditional forms of rural energy in two main ways: firstly, it replaces fossil fuels with clean methane, which reduces not only the release of greenhouse gases but also other harmful emissions; secondly, the various forms of anaerobic digestion facilitate the more efficient use of organic waste (Collet et al. 2011; Rehl and Müller, 2011). Such procedures are of particular importance in dealing with the growing pressure of problems related to global energy scarcity and climate change (Chen et al. 2012).

2.3 Biogas combustion process

In the combustion process, the fuel must be mixed with an oxidizer for combustion to take place. The complete combustion reaction of methane, contained in biogas, with oxygen is (Biogasburner, 2011):

$$CH_4 + 2O_2 \rightarrow CO_2 + 2H_2O \quad (1)$$

One volume of methane requires two volumes of oxygen to produce one volume of carbon dioxide and two volumes of water vapor, assuming that there is 58 % methane in the biogas and 21 % oxygen in the air (Souza, 2011).

In direct combustion, biogas is burned in the combustion chambers of gas turbines, boilers, heaters and dryers. The heat released during burning is used in production processes to generate electricity. It can also be converted into mechanical or electrical power in internal combustion engines. Internal combustion engines coupled to electric generators, called motor generators, can be used on rural and agro-industrial properties where residual biomass and therefore biogas is available for the production of electricity.

2.4 Lower calorific value

The lower calorific value (LCV) is used to determine the theoretical potential energy contained in fuels. It varies according to the concentration of methane. The higher the calorific value of biogas (PCI), the higher the concentration of methane in the biogas. Table 1 shows the lower calorific value of some fuels. The specific weight and density of biogas depend directly on the methane concentration.

Table 1 - Lower calorific value of biogas depending on the chemical composition of the biogas

Chemical composition of biogas	Specific weight $(kg\ m)^{-3}$	Lower calorific value $(kcal\ kg)^{-1}$
10% CH_4 and 90% CO_2	1, 8393	465,43
40% CH4 and 60% co2	1,46	2333,85
60% CH4 and 40% co2	1,2143	4229,98
65%CH4 and 35% co2	1,1518	4831,14
75% CH4 and 25% co2	1,0268	6253,01
95%CH4 and 5% co2	0,7768	10469,60
99% CH4 and 1% co2	0,7268	11661,02

SOURCE: IANNICELLI (2008).

2.5 Benefits of biogas

Considering the environmental benefits, the use of biogas has been defended by scholars due to the fact that it pollutes the atmosphere less with lower GHG emissions, helping to reduce global warming. Not to mention the reduction in deforestation due to the replacement of firewood with biogas.

From an agricultural point of view, fossil fuel derivatives are replaced, saving on transportation costs. Solid waste is also used as organic fertilizer: biofertilizer. Biogas contributes to improving the quality of life in rural areas. Even today, the use of wood-burning stoves and diesel-powered tractors is common. It is hoped that with the increased use of biogas, these fossil fuels will be replaced.

The use of biodigesters has grown considerably in recent years. This is because, in addition to generating biogas and biofertilizer, it solves the problem of treating waste, whether domestic or animal, and contributes to maintaining an economically sustainable environment.

2.6 **Aspects of bioenergy**

Bioenergy refers to the energy obtained from biomass, which is the biodegradable fraction of products and residues from agriculture (of plant and animal origin), forestry and related industries, as well as the biodegradable fraction of industrial and urban waste (FAO, 2008). A wide variety of biomass sources can be used to produce bioenergy in various ways. Residues from food processes, fibers and wood, the industrial sector, energy crops, agricultural and forestry residues can be used to generate electricity, heat, combined heat and power and other forms of bioenergy (GBEP, 2007).

Bioenergy production and trade have increased significantly over the last few years and are projected to rise even further over the next few decades. According to the International Energy Agency, world biofuel production will rise from 0.8 EJ in 2005 to 2.3 EJ in 2015 and by 2030 it is estimated at 3.9 EJ (IEA, 2006). However, concerns have been raised about the potential negative impacts of bioenergy production, such as the impact on food security in developing countries and the impact on biodiversity.

Business involving the bioenergy trade has expanded rapidly in recent years. Forestry and agricultural residues, wood, pellets and briquettes, liquid biofuels (vegetable oils, bio-ethanol, biodiesel) are increasingly being marketed for applications in the production of electricity, residential heating and fuel for means of transport. The driving force behind the expansion of bioenergy is a renewable and accessible source that contributes to climate change mitigation, energy security and rural development (Junginger et al. 2008; Erb et al. 2012).

Large-scale bioenergy production systems are evaluated according to sustainability criteria that take into account social, environmental and economic impacts. National and supranational initiatives are being developed and implemented to encourage bioenergy production and at the same time certify projects in terms of sustainability (Dam et al. 2008).

The Brazilian government has promoted the expansion of bioenergy, with a focus on ethanol and biodiesel, through programs and laws: the National Alcohol Program (PROÀLCOOL, Decree 76.593/75), the Innovation Law (Law 10.973/04), the Biodiesel Law (Law 11.097/05), the National

Agroenergy Plan (PNA 2006-2011), the Low Carbon Agriculture Program (ABC Program), the Program to Encourage Alternative Sources of Electricity (PROINFA, Decree 5.025/2004) and the National Program for the Production and Use of Biodiesel (PNPB), to stimulate the participation of biofuels in the national energy matrix (Duraes, 2008). Brazil is in a position to lead the world in the production of bioenergy, due to its extensive agricultural land and geoclimatic conditions, because part of these resources are unexploited or underutilized (Goes et al. 2010; Martha JR, 2008).

3. CONCLUSIONS

This chapter briefly reviews the literature on alternative energies, focusing on biomass and biogas. With the need to renew the country's energy matrix, the federal government has been implementing more and more programs to encourage the production of renewable energies, with a view to environmental sustainability in the country's energy production.

Chapter II - Financial analysis of electricity micro-generation using pig biogas

SUMMARY

One of the biggest sources of energy available in rural and agro-industrial areas is biomass. It comes in the form of plant and animal waste, such as crop residues, animal manure, energy plants and agro-industrial effluents. This waste can be used by the farmer or agro-industry to burn directly to produce heat or to produce biogas in biodigesters. Pig production generates a large amount of waste, causing problems for the environment when left untreated. It contains a large amount of methane which, when released into the atmosphere, contributes significantly to the greenhouse effect. Incentive programs for producers have been created to preserve the environment by minimizing the effect of gases on nature. In this context, the Incentive Program for Alternative Energy Sources (Proinfa) and the ABC Program were created on the basis of the Kyoto Protocol, with the aim of encouraging farmers to invest in sustainable food production. The co-generation of electricity is one of the ways of using the biogas generated by food production, mainly meat products. As well as generating energy, there is also the possibility of selling carbon credits, adding to the producer's income. The aim of this chapter is to determine the economic viability of co-generating electricity with biogas from pig waste, adding value by selling carbon credits. This case study shows that the cost of producing electricity depends directly on the initial investment and the time it takes to generate the energy. Under current biogas production conditions, the cost of a MWh is R$ 289.94 and the value of a m^3 of biogas is R$ 0.25. With interest of 5.5% p.a., the producer would lose around R$ 224,739.21. Considering the sale of carbon credits would still generate losses, but smaller ones of around R$ 18,519.21 and the cost of producing energy per MWh is R$ 111.05.

1. INTRODUCTION

For the next few decades, *there are* prospects of a crisis in the energy sector. This crisis is due to the mismatch between the growth in demand and the inability of supply to keep pace with the expansion of the world's Gross Domestic Product (GDP), especially oil, which is the basis of the national energy matrix. The immediate and perceptible result of this crisis is volatility and record highs in the price of crude oil. Countries are trying to alleviate the uncertainty and somehow prevent their economies from collapsing in the face of the energy crisis by stimulating the use of renewable energy sources such as biomass, solar and wind. (Castro, 2008).

With the rise in the value of conventional energy sources, together with growing concern about the future of energy supply, energy security has gained an important position in political circles around the world, including Brazil. Recognized for its renewable energy sources, the country has become the focus of attention in developed countries, in addition to hydroelectric plants, the use of clean energy sources such as biofuels.

The financial aspect is often cited as the main motivation for adopting new energy sources. This is because the cost of exporting oil to countries that don't have this raw material becomes high. Environmental and social concerns have become important, given that globalization has had devastating effects, as it has accelerated economic growth and, in turn, generated interdependence between markets, the fruits of which are not enjoyed by the entire population. Poverty still persists, and disrespect for the environment is beginning to have its consequences, such as a decrease in biodiversity and an increase in global warming.

The west of Paranà is the largest pig-producing region in the state, generating a serious environmental problem. Untreated waste pollutes rivers and releases gases that increase the greenhouse effect. Biogas from pig farming is becoming an important source of renewable energy. As well as producing electricity to meet producers' needs, it contributes to reducing environmental damage. However, there are aspects of this production that need to be analyzed carefully, such as the financial aspect of implementing projects to produce electricity.

In order for sustainable food production to become commonplace among producers, it is necessary to study greenhouse gas emissions, climate change and the instability of energy prices. Agriculture is one of the most important activities and although it produces food, it can, if poorly conducted, pollute rivers and springs, promote the removal of forests and contribute to GHG emissions (through animal production), among other aggravating factors.

With the expansion of pig production in Brazil, the amount of waste generated has also increased, causing a huge environmental problem. Biogas production has become an alternative for reducing GHG emissions, and is being encouraged by the country's private and public sector. Solutions are being implemented to reduce environmental pollution, such as stabilization ponds and biodigesters. They treat waste and turn it into biogas and biofertilizer.

This chapter aims to show the financial viability of producing electricity from pig biogas.

2. LITERATURE review

2.1 Rural waste as a source of energy

Brazil is a country with a tradition of using renewable energy sources, with hydroelectric power accounting for the largest share (74%) of all electricity generation. On the other hand, there is still a huge potential for renewable energy sources, including wind energy (0.4%) and biomass (4.7%) (BEN, 2011).

According to data from ABIPECS (2012), pig production and slaughter grew dramatically between

2006 and 2010. In 2006, 23,134,111 head were slaughtered. In 2010, 29,072,584 head were slaughtered, a considerable increase in production over the five-year period. The production of waste has increased, so care has to be taken to treat it. According to data provided by the IBGE (2006), Brazil is the world's fourth largest producer of pigs, with around 43.2 million head of pigs.

Rural waste includes all types of waste generated by productive activities in rural areas, namely agricultural, forestry and livestock waste. Livestock waste consists of manure and other products resulting from the biological activity of cattle, pigs, goats and others, whose local relevance justifies their energy use. The production of biogas from manure contributes to environmental protection, reducing CO2 emissions by replacing fossil fuels and reducing methane (CH4) emissions (Moller et al. 2007).

In addition to the biogas produced, anaerobic digestion transforms the raw material added to the biodigester into fertilizer for agricultural production. The degradation process increases the availability of nitrogen to the plant, thus increasing the fertilization efficiency of the raw material (Borjesson and Berglund, 2003, 2005).

In rural areas, alternative renewable energy sources can be used in isolated rural communities to improve their living conditions. Implementation depends on the availability of energy resources in each region. In places where animal waste is available and cannot be disposed of directly in nature before undergoing a treatment process, biogas becomes available. Biogas is a by-product of the anaerobic digestion of animal waste and can be used as a biofuel to heat rooms, cook food and in motor generator sets to generate electricity.

2.2 Renewable energy sources

The replacement of fossil fuels, the introduction of measures to increase the efficiency of energy use and conversion and heavy investment in the development of renewable energy sources (solar, wind and others) and the production of "clean" fuels (derived from biomass and hydrogen) were solutions found by the countries belonging to the IEA at various international discussions (Rio-92, Kyoto-97 and Bonn-2001) to make it possible to reduce GHG emissions without reducing the quality of life provided by the use of non-renewable energy sources (Silva et al. 2003).

The generation of energy using pig waste is still remote in Brazil, even though it is a way of reducing pollution and environmental degradation around farms. However, the current system is very favorable to development, due to the cycle of high prices for oil, fossil fuels and natural gas - making this type of generation viable, not to mention the growing pressure to reduce environmental degradation associated with food production (Lemos et al. 2008). With the prospect of selling carbon credits, there has been growing interest and demand for biodigesters from pig producers since 2005 (MCTI, 2007).

This has led to an increase in the availability of biogas and the emergence of electricity generation, both for their own consumption and for sale to utilities (Martins and Oliveira, 2011).

Biomass was the main source of energy until the beginning of the 20th century, when the so-called "oil age" came along and biomass energy was relegated and largely ignored. Biomass is generally referred to as the total mass of organic matter that accumulates in a living space. The use of biomass to generate electricity has been the subject of several studies and applications, both in developed and developing countries. Among other reasons are the search for more competitive sources of energy generation and the need to reduce carbon dioxide emissions (Silva et al. 2005).

Biomass waste is considered an available source of energy, although it can cause serious environmental degradation. The recovery of biogas through anaerobic digestion is seen as the ideal way to treat biomass waste. Different types of waste have the potential to produce biogas, such as cattle waste, pig waste, sewage sludge, fruit and vegetable waste, among others (Qiao et al. 2011).

Biogas produced through anaerobic digestion is basically composed of methane (CH4), carbon dioxide (CO2) and hydrogen sulphide (H2S). Anaerobic digestion is a natural process in which anaerobic bacteria attack the structures of organic matter to produce simple compounds such as methane, carbon dioxide and water, thus forming biogas. Biogas contains approximately 36 to 50% methane (CH4) and 15 to 60% carbon dioxide (CO2) (Kao et al. 2012; Starr et al. 2012; Richebosh et al. 2011).

Biogas is a by-product generated by the process of treating pig waste, which, when used, allows rural producers to become self-sufficient in electricity and pay for the capital invested in a waste treatment system. According to Souza et al. (2004) another advantage of using biogas is that methane is a gas that contributes more to the greenhouse effect than carbon dioxide and burning it during the energy generation process helps to reduce its effect as such.

The construction and operation of plants that produce energy from renewable sources is the subject of discussion in all countries that have accepted the concept of sustainable development and the countries that have signed the Kyoto Protocol aimed at sustainability. The expansion of renewable energy production is clearly an imperative, but its construction is only economically viable if its operation results in long-term sustainability (Gulosin et al. 2012).

According to Mohseni et al. (2012), renewable energy (mainly wind and solar) has gained a lot of attention in recent years and will play an important role among future primary energy sources. According to data from the MME (2007), the supply of electricity from renewable sources is set to increase significantly. Projections for the coming years show that energy from sugar cane will increase in 2030 compared to 2010, when production will reach 103,026 Mtoe.

The use of biomass gas to generate electricity aims to reduce energy production costs, reduce GHG emissions and increase the supply of electricity in Brazil (MME, 2007). Lindemeyer (2008) says that a positive aspect of generating energy from biogas is that it has great potential to boost the local economy, encouraging important sectors such as industry, commerce and services.

Figure 1 summarizes the possibilities for using biogas as an alternative energy source.

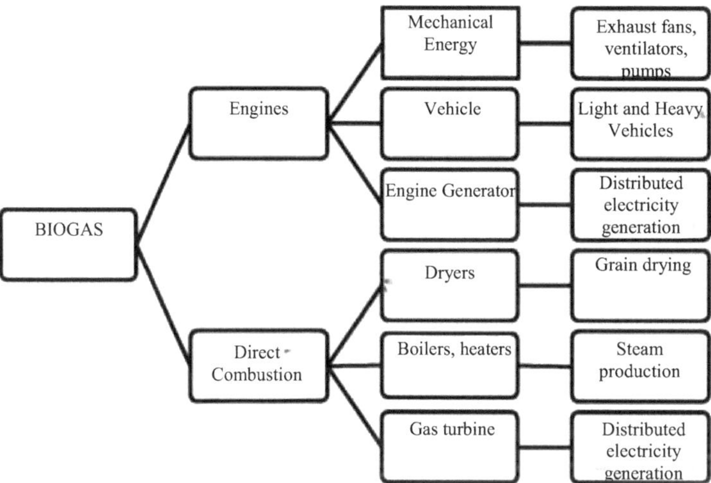

Figure 1 - Flowchart of biogas use. Source: Walsh et al. (1988) and Center for Energy Conservation (CCE) (2000).

2.3 Legislation and incentive programs for electricity production

According to ANEEL (2002), distributed energy generation in Brazil has the following purposes:

Ensure the expansion of supply and quality of supply without increasing prices, based on competition and stimulating the economic efficiency of agents;

Release the taxpayer from future investments, stop the generation of liabilities and recover resources through privatizations;

Serving the citizen allows the Government to be more efficient and focused on the Social Field: Health, Education and Security.

Distributed generation is still little explored, but it has great potential to diversify the energy matrix, reducing environmental impacts and boosting the development of industry by generating jobs and income (ANEEL, 2002).

They were officially defined by Decree No. 5.163 of July 30, 2004, as follows:

Art.14. For the purposes of this Decree, distributed generation is considered to be the production of electricity from projects by concessionaires, permit holders or authorized agents (...), connected directly to the buyer's electrical distribution system, except for that from a project:

I hydroelectric plant with an installed capacity of more than 30 MW; and

II thermoelectric, including cogeneration, with an energy efficiency of less than seventy percent, (...).

Sole Paragraph. Thermoelectric projects that use biomass or process waste as fuel will not be limited to the percentage of energy efficiency provided for in item II of this heading.

The PRODIST (Distribution Procedures) definition of distributed generation is the generation of electricity, of any power, connected directly to the electrical distribution system or through consumer installations, which can operate in parallel or in isolation (ANEEL, 2005).

As a reference in this work, distributed generation is considered to be any generating source with output destined, for the most part, for local loads, fed without the need to transport the energy through the transmission network and with the capacity for mobility in terms of its physical location (INEE, 2002).

Technologies related to distributed generation are related to renewable energy cogeneration systems, such as wind generation, photovoltaic systems, small hydroelectric plants and the use of biomass to produce electricity. Generally, these sources are located close to consumer centers, thus reducing energy transmission and distribution costs (Januzzi, 2000).

The Ministry of Mines and Energy and the Electricity Crisis Management Chamber (GCE) have drawn up a specific incentive program with the aim of adding 2,000 MW of biomass generation to the national electricity system. *There are* also specific financing programs with reduced interest rates and grace periods compatible with the project (ANEEL, 2008).

Studies from the Atlas of Electricity in Brazil (ANEEL, 2008) show the cost of producing energy from various sources, both renewable and non-renewable. The value of electricity produced from biomass is the lowest, at around R$101.75 MWh $.^{-1}$

The amounts paid in the 2nd Renewable Sources auction, including wind power, SHPs and biomass, are R$130.86, R$141.93 and R$144.20 respectively (CCEE, 2010). The aim of these auctions is to encourage the production of renewable energy, making it more profitable for producers.

The reduction of greenhouse gas emissions came about with the Kyoto Protocol signed in 1997 in the city of Kyoto (Japan), which instituted the reduction of GHG rates in the atmosphere. For the Protocol to be effective, at least 55 countries had to ratify it, which together should account for 55% of global GHG emissions. In view of the economic, social and development differences, the countries were divided into two groups:

Annex I: These are the developed, industrialized countries and some countries with economies in transition, including Russia and Eastern Europe, which are considered to be the biggest GHG emitters;

Not Annex I: Composed of developing countries including Brazil, Argentina, Mexico and India.

The countries included in Annex I should reduce their combined GHG emissions by at least 5.2% compared to 1990 levels by 2012. Japan and the European Union have made larger commitments, 6% and 8% respectively. In order for countries to achieve their reduction targets, the Protocol stipulated three flexibility mechanisms, the first two of which are only valid for Annex I countries:

Joint Implementation: this means that Annex I countries can act together to achieve their targets. In other words, if a country doesn't manage to reduce its emissions sufficiently, it can enter into an agreement with other Annex I countries to help each other;

Emissions trading: which takes place between Annex I countries, in which companies from countries with emissions below their ceiling in the period between 2008 and 2012 would have CERs that could be sold to companies from countries with emissions above their ceiling;

Clean Development Mechanism - CDM: allows activities between Annex I and Annex II countries, with the aim of supporting sustainable development (Viola, 2009).

According to Key and Stacy (2011), an offset market for pig producers helps to reduce greenhouse gas emissions, based on carbon credit payments. The factors that influence the production of biogas using biodigesters and that determine the characteristics and location of pig activities can facilitate the introduction of a carbon offset market. These include the size of the operation, the selling price of surplus electricity, the economic incentive, the level of methane emissions and the price per ton of carbon.

Brazil has advanced environmental legislation, whose application in terms of regulatory instruments favors "command and control" measures, such as environmental licenses, limits on the emission of pollutants and zones where certain activities are prohibited or restricted due to potential environmental damage.

PROINFA (Incentive Program for Alternative Energy Sources), created on the basis of Law No. 10.438/02 (MME, 2005), aims to increase the share of electricity by adding 3,300 MW of installed capacity to the electricity system, produced by Independent Autonomous Producers from wind power, small hydroelectric plants (SHPs) and biomass in the national interconnected system. This opens up an opportunity for electricity generation systems to be set up using biogas as their primary energy source, thereby promoting the participation of this renewable source as an alternative energy source in the national energy matrix.

PROINFA's incentive mechanisms are the guaranteed purchase of energy for a period of up to 15 years and the establishment of a reference value compatible with the technical and economic characteristics of the project (ANEEL, 2008).

In 2010, the Low Carbon Agriculture Program (ABC Program) was created, which provides resources and incentives for producers to produce food in a sustainable way, contributing to the reduction of greenhouse gas emissions (CO_2, CI4, and nitrous oxide). The main idea behind the program is for producers to increase their income, increase production and protect the environment. It is expected that for the 2011/2012 harvest, the ABC program will provide approximately R$3 billion to encourage technological processes that minimize or neutralize GHG emissions in the field (Brasil, 2012).

The Distributed Generation with Environmental Sanitation Program implemented by Copel (2009) in the state of Paranâ aims to encourage the installation of equipment that produces electricity for both own consumption and the sale of surplus energy, as well as minimizing the effects caused by animal waste and agro-industrial waste on the environment. The program encourages the sustainable elimination of the causes of pollution resulting from the discharge of animal waste into rivers and reservoirs. In the state of Paranà, 6 pilot distributed generation projects are underway, including rural producers and agro-industries. The pilot projects total 524 kW, enough to supply 130 households.

2.4 Financial scenario for electricity production

Studies on the generation of electricity from biogas generated by pig manure are recent (Oliveira and Higarashi, 2006; Zago, 2003), and the inclusion of economic analysis is still little explored. Singh and Sooch (2004), studying the cost of installing biogas plants with cattle manure, found that as the biogas plant increases, the cost of installation and the operating cost increase proportionally, regardless of the type of biodigester installed.

With the prospect of trading carbon credits, there has been a growing interest and increase in the adoption of biodigesters by pig producers since 2005 (MCTI, 2007). Waste from animal husbandry such as poultry and pig farming, for example, has a high potential for pollution. Traditionally, they have been dumped directly into the soil as fertilizer, but in some situations these methods can cause environmental problems, such as unpleasant odours and water contamination. Two options for energy use have been studied for this type of waste in Europe: one is the anaerobic digestion of pig waste, and the other is direct combustion, using poultry litter (Souza et al. 2004).

Burning 1 ton of methane is equivalent to removing 21 tons of carbon (IPCC, 2007). For Lazzarus et al. (2011) the destruction of methane can provide a source of income, because burning methane provides credits to be traded. These credits, usually measured in terms of CO2, become a complementary source of income. According to Liebrand and Ling (2008), credits traded on the Chicago Climate Exchange (CCX) ranged from US$1.90 per ton to US$7.40 in 2008, with an average price of US$4.98 per ton of carbon. Data updated by the ICB (2012) shows that at the end of February 2012, the price paid per ton for delivery in 2013 reached US$14.50.

2.5 Biodigesters

According to Fernandes (2012), biodigesters are designed to contain residual biomass that comes into contact with microorganisms, in the absence of oxygen, where there is the production and preliminary storage of gaseous compounds or by-products of the process such as biogas and biofertilizer.

The change of organic compounds into simpler compounds takes place in stages inside the biodigester: first, complex organic molecules are broken down into simple, soluble ones through hydrolysis, followed by acidogenesis, which transforms glucose, amino acids and fatty acid molecules into organic acids, alcohols and ketones, which are transformed into acetate, carbon dioxide and hydrogen. In the last phase, methanogenesis takes place, producing methane (Marques, 2012).

Biodigesters have received the most attention because of their ability to reduce greenhouse gas emissions from waste treatment. Sedimentation ponds and other manure treatment systems emit greater quantities of methane.

The most widely used type of biodigester in Brazil is the Canadian biodigester. According to Deublein and Steinhauser (2008), it is a horizontal model, with a masonry loading box that is wider than it is deep, and therefore has a greater area of exposure to the sun, which allows for high biogas production and prevents clogging. During biogas production, the biodigester dome inflates because it is made of soft plastic material (PVC) and can be removed.

3. MATERIALS AND METHODS

3.1 Characteristics of the Study Area

The Granja Colombari unit is located on Linha Marfim, in the municipality of Sao Miguel do Iguaçu, latitude 25°20'53 South and longitude 54°14'16 West, in the west of the state of Paranà. It has a total area of 250 hectares, of which 200 hectares are used for agricultural production and 50 hectares for pig farming, Permanent Preservation Area (APP) and Legal Reserve. Since 1997, it has been dedicated to raising pigs in the finishing phase, with a significant increase in pig production over the years.

The Colombari farm works with a confined rearing system. Biomass generation is directly linked to management factors, the water supply system, air conditioning and cleaning systems. The property has two biodigesters.

Figure 2 - Biodigesters at the Colombari Farm (SOURCE: Marques, 2012)

The first has dimensions of 25m x 10m x 3.7m and a capacity of 29.2 m^3 d-1, with a water retention time of 30 days. The waste then goes to the second biodigester with a capacity of 7.3 m3 d-1, designed for a water retention time of 20 days. The biodigesters have an internal pressure relief valve: a water box, with a pipe coming out of the biodigester, inserted 15 mm deep into the water, so that at pressures above 15 mm of water column, the biogas is released from the biodigester. At the outlet pipe, the effluent is directed to a manure house that stores the biofertilizer, which is then used on the crops.

Electricity is generated by the motor-generator unit, the protection and control system and the control panel, which is interconnected to the distribution network to sell the surplus. The motor-generator is housed in a masonry house with a non-expanding reinforced concrete floor and air inlets for internal ventilation. The technical data for the engine and generator are shown in Table 1 and Table 2.

Table 1 - MWM 6.12T Biogas Engine - Diesel Cycle converted to Otto Cycle

Engine capacity	7.2 L
Ignition	Electronics (Pandoo)
Rotation	1800 RPM
Biogas consumption	50 m^3 h^{-1}

Table 2 - Generator: Gramaco / G2R 200MB 4

Number	17882
Number of poles	4
Nominal speed	1800 RPM

Power	104 kVA continuous
Frequency	60 Hz
Voltage	127 V
Current	270 A
Cos φ	0.8 a 1.0
Insulation class	H
Manufacturing	2010

Figure 3 shows the engine coupled to the 100 KVA generator.

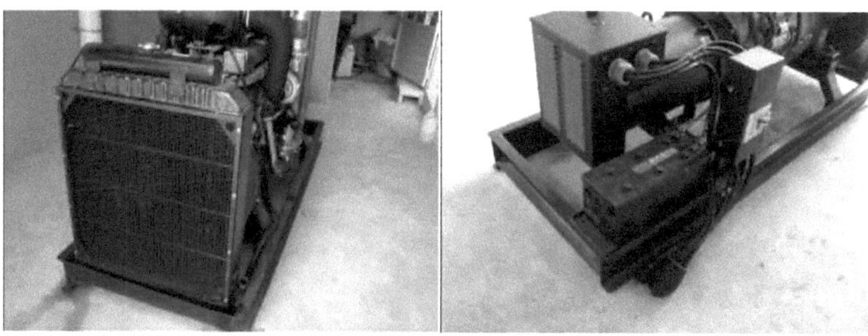

Figure 3 - Engine coupled to the generator (Cleber, 2012)

The financial analysis varied the time it took to generate electricity and, depending on the number of pigs, the initial investment, including the sale or not of carbon credits. The cost of installing the biodigester and the motor-generator was obtained from companies in the sector.

3.2 Equations for determining financial parameters and the cost of producing electricity.

Three scenarios were studied.

Scenario 1 - Colombari Farm's current situation. The average biogas production is 554 m d^{3-1} , for an average of 4673 pigs housed at an average temperature of 22.11°C. The motor-generator data provided by the manufacturer is: power - 76 kW and consumption of 50 m^3 h^{-1} of biogas. The data collected from the operation of the motor-generator is an average power of 66.22 kW and a specific consumption of 45.55 m^3 h^{-1} of biogas. Actual operating data will be used for calculation purposes. Energy generation runs for 10 hours a day. The cost of the biodigester is R$294,600.00 and the motor generator R$113,550.00.

Scenario 2 - Simulating energy generation for 16 hours a day. For this, biogas production will be 729 m3 of^{1} and simultaneously pig production will reach 6073 head. The power and consumption of the motor will be the same in all three scenarios. The cost of the biodigester will be R$ 352,800.00. The cost of the motor generator for calculation purposes will be R$ 113,550.00.

Scenario 3 - Simulating 20 hours of energy generation per day, with 911 m3 of[1] biogas and a pig production of 7000 animals. The cost of the biodigester is R$ 441,000.00 and the motor generator is R$ 113,550.00.

The commercial values for the biodigester (including excavation and complete installation) were supplied by Sansuy - Soluçoes em PVC. The values for the motor-generator were supplied by ER-BR Energias Renovâveis.

The cost of electricity production will be calculated in all three scenarios. The parameters for the economic evaluation of projects are defined below.

Net Present Value (NPV) - compares the investment made with the present value of the cash flows generated by the project. It considers all the cash flows, not just the instant in time when the accumulated balance becomes positive. It can give us a measure of wealth added (NPV > 0) or destroyed (NPV < 0). The interest rate to be used to define the parameters will be 5.5% p.a. as imposed by the ABC Program. NPV is given by Bordeux-Rêgo et al.(2010):

$$VPL = -I + \left(\sum_{t=1}^{n} \frac{FC_t}{(1+i)^t} \right) + \frac{VR}{(1+i)^n} \tag{2}$$

Where:

I - Initial investment;

FCt - Net cash flow on date t;

i - interest rate;

RV - Residual value of the project at the end of the analysis period; n - investment amortization time.

The equations used for the financial analysis were adapted from Souza et al. (2004).

Internal Rate of Return (IRR) - This is a benchmark to be used to define whether or not a project will be accepted. The internal rate of return makes the NPV zero.

Time to return on investment (TRI) - for the calculation of TRI, the value paid to the producer for the MWh produced was simulated at R$120.00 - R$140.00 - R$160.00 - R$180.00 - R$200.00 - R$220.00, because according to ANEEL (2010) the value paid for energy from biomass is R$144.20 MWh^{-1} . For the feasibility analysis, the value considered will be R$ 140.00 MWh .$^{-1}$

$$TRI = \frac{\ln\left(\frac{k}{k\times(1-i)^{-1}-i}\right)}{\ln(1+i)} \quad (3)$$

Where:

$$k = \frac{A}{CI} - \frac{OM}{100}$$

CI - Investment cost of the biodigester/motor-generator system (R$), A - Annual expenditure on electricity purchased from the grid (R$ per year^{-1}), OM - Expenditure on amortization and maintenance of the plant (R$ per year^{-1}), TRI - Payback time (years).

Cost of electricity:

$$Ce = \frac{CAG + CABi}{PE} \quad (4)$$

in which

Ce - Cost of electricity produced via biogas (R$ kWh^{-1}), CABi - Annual expenditure on biogas (R$ year^{-1}) and PE - Production of electricity by the biogas plant (kWh year^{1}), CAG - Annualized cost of investment in the motor-generator set (R$ year^{-1}).

Where:

$$CAG = CIG \times FRC + \frac{CIG \times OM}{100} \quad (5)$$

$$CAB = CB \times CNB \quad (6)$$

in which:

CIG - Cost of investment in the motor generator (R$), OM - Cost of organization and maintenance (%/year), CB - Cost of biogas (R$$^{m-3}$) and CNB - Consumption of biogas by the motor generator set (m^3 year).$^{-1}$

The production of electricity (PE) is given by:

$$PE = Pot \times T \quad (7)$$

Pot - Nominal plant power (kW), T - Annual plant availability (h year^{-1}). The capital recovery factor is given by

$$FRC = \frac{i(1+i)^n}{(1+i)^{n-1}-1} \quad (8)$$

FRC - Capital Recovery Factor, i - discount rate (% per year) and n - years to amortize the investment.

The cost of biogas is given by

$$CB = \frac{CAB}{PAB} \quad (9)$$

CAB - Annualized cost of investment in the biodigester (R$ per year^{-1}) and PAB - Annual production of biogas (m^3 per year).$^{-1}$

$$CAB = CIB \times FRC + \frac{CIB}{100} - CC \quad (10)$$

CIB - Biodigester investment cost (R$) e PAB - Annual biogas production (m^3) - GCC - Carbon credit gain (R$ year)$^{-1}$

To convert the amount of methane into carbon credits, equation (11) was used.

Conversion of biogas into tons of CO_2 and converted into R$ t .$^{-1}$

The following parameters were used for this conversion:

Biogas production in m^3 year^{-1} (Pr)

S Methane density: 0.67 kg at 20°C and 1 atm pressure (Angonese et al. (2007); Honório (2009); Pukasiewicz (2010))

s Percentage of methane in biogas: 60%; (%CH)$_4$

s Value per ton of CO_2 : R$25.37

s Equivalence between CH_4 and CO_2 - 1 unit of CH_4 equals 21 units of CO_2 .(Un)

$$t\ CO_2=(Pr \times Ds \times Pc \times Un)/1000 \quad (11)$$

$$valor\ da\ t = t\ CO_2 \times R\$ \quad (12)$$

1. 6 **Profitability index**

The profitability index is a relative measure between the present value of the cash flows received and the initial investment according to Bordeaux-Reo (2010), given by equation (13).

$$IL = \frac{VPL+I}{I} \quad (13)$$

Where,

IL - Profitability Index; NPV - Net Present Value; I - Investment

Initial.

When IL is greater than one (1), it means that the investment will be recovered, remunerated at least at the required rate and there will also be an increase in wealth.

When IL is equal to one (1), the investment will be recovered, remunerated at exactly the required rate.

When IL is less than one (1), the investment will not be recovered, resulting in losses.

4. RESULTS AND DISCUSSION

4.1Scenario 1

Figure 4 shows the variation in the return on investment as a function of the hours of energy production and the amount paid per MWh. With production of 10 hours^{-1} without the sale of carbon credits, the payback time is around 35 years. With the revenue from the sale of carbon credits, the TRI falls to 8.5 years, considering the sale price of R$140.00 MWh .$^{-1}$

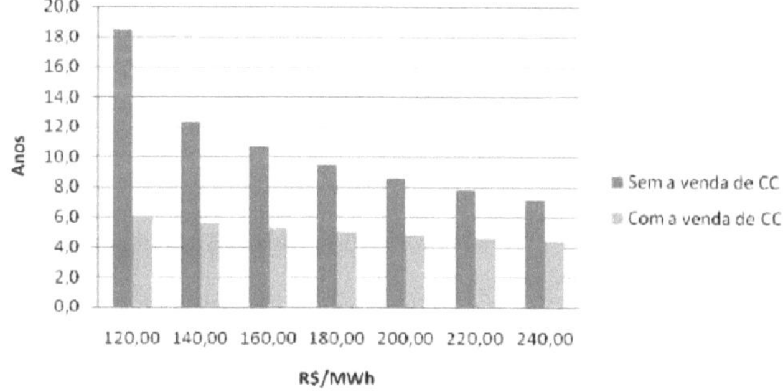

Figure 4 - Return on Investment (ROI) as a function of the value of electricity - 10h d^{-1}

Table 3 shows the variation in the cost of electricity according to whether or not carbon credits are sold. Producing energy for 10h d^{-1} without selling carbon credits, with a return on investment estimated at 10, 15 or 20 years, the value of energy production is R$289.94, R$219.58 and R$185.80 per MWh respectively, higher than the amount paid by the energy concessionaire, which is R$140.00 per MWh.

Table 3 - Cost of electricity (R$.MWh^{-1}) from pig biogas, varying the production time as a function of the useful life of the biodigester.

Biodigester lifespan (years)	10h d^{-1}	
	No sale of carbon credits	With the sale of carbon credits
10	R$ 289,94	R$ 111,05
15	R$ 219,58	R$ 40,65
20	R$ 185,81	R$ 6,86

4.2 Scenario 2

With production at 16h d^{-1} , the IRR without the sale of carbon credits is 18.5 years. With the sale of carbon credits, the payback time is 6 years. As the amount paid for surplus production increases, the payback time decreases.

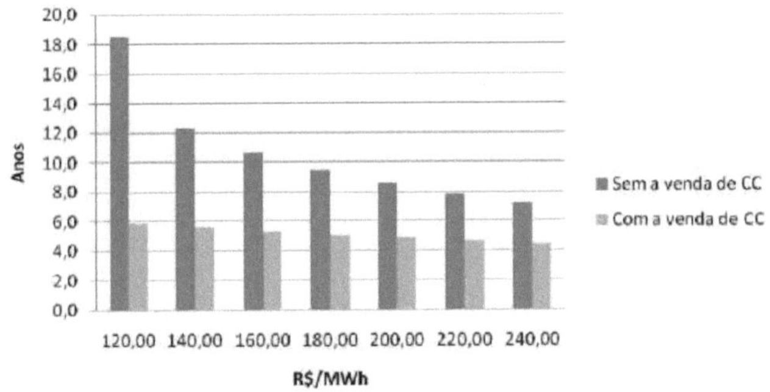

Figure 5- Return on Investment (ROI) as a function of the value of electricity - 16 h day^{-1}

Oliveira & Martins (2007) studied the economic viability of using a generator providing 40 kWh, using biogas from pig farming, and concluded that this alternative is viable as demand and the price of energy increase. With a tariff of R$ 200.00 MWh^{-1} , the return on investment, considering a discounted interest rate of 5.5% p.a. (ABC Program), was 39, 26 and 19 months for daily generation times of 10, 14 and 18 hours, respectively.

Table 4 shows that with 16 h d^{-1} , the value becomes competitive after the 20-year payback period when the value of energy production reaches R$ 132.60 MWh^{-1} . Considering the sale of carbon credits, with 10 years it becomes feasible when the cost of producing electricity is around R$ 59.88, generating extra income for the producer by selling the surplus energy to the utility company.

Table 4 - Cost of electricity (R$.$MWh^{-1}$) from pig biogas varying the production time as a function of the useful life of the biodigester

Biodigester lifespan (years)	16 h.d^{-1}

	No sale of carbon credits	With the sale of carbon credits
10	R$ 207,10	R$ 59,88
15	R$ 156,70	R$ 9,58
20	R$ 132,60	R$ -14,54

4.3 Scenario 3

When energy production is 20 h d^{-1} , the IRR without the sale of carbon credits is 14.5 years. With the sale of carbon credits, it is approximately 6 years.

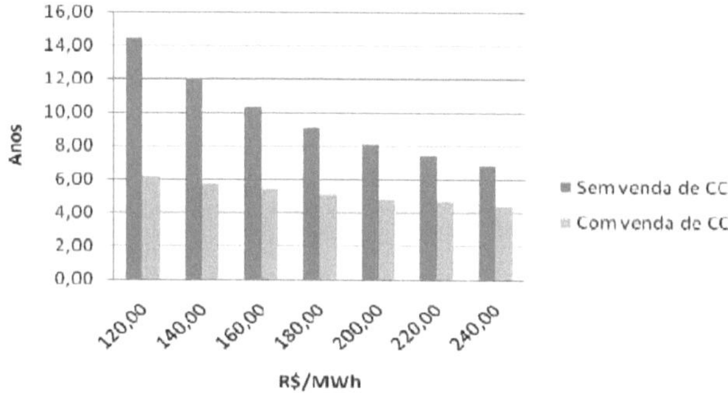

Figure 6 - Payback time (TRI) as a function of the value of electricity - 20 h d^{-1}

Table 5 shows the costs of producing electricity for 20 hours of operation^{-1} , varying the return on investment time. It can be seen that, without the sale of carbon credits, the cost of producing energy becomes competitive considering the 20-year IRR. With the sale of carbon credits, the cost of production is low (R$ 49.80 MWh^{-1}). With a 20-year TRI, there is no cost to produce energy, which is evident from the fact that the production cost is negative.

Table 5 - Cost of electricity (R$.MWh^{-1}) from pig biogas with production time as a function of the useful life of the biodegenerator

Biodigester lifespan	20 h.day^{-1}	
(years)	No sale of carbon credits	With the sale of carbon credits
10	R$ 197,02	R$ 49,80
15	R$ 149,10	R$ 1,95
20	R$ 126,23	R$ -21,00

According to Zago (2003), the project becomes economically viable when the property has a production capacity of 200 m^3 .d^{-1} of gas, which generates an approximate production of 300 kVAh.d$^-$

[1] . For Stegelin (2010), the benefits of energy generated by biogas consist of savings on the cost of electricity purchased, savings on fuel when biogas is used for this purpose; savings on the purchase of natural gas for heating water and other facilities; avoiding the cost of commercial fertilizers by using raw manure for fertilization, plus the revenue from carbon credits on the greenhouse gas markets, confirming the results obtained.

Yiridoe et al. (2009) and Bronw et al. (2007), studying the feasibility of producing biogas to generate electricity in dairies, found that it was not viable (US$0.12.kWh^{-1}) to produce electricity in small dairies. For small pig producers, biogas production has become unfeasible, at the same level as for small dairies (US$0.12.kWh). Gwavuya et al. (2012) studying the use of biogas in households in Ethiopia found a rate of return close to 10% with a biogas plant with a capacity of 6 m^3 , using animal waste.

Coldebella et al. (2006) analyzed the production of electricity via cattle biogas and found that the cost of producing electricity is directly associated with the time it takes to amortize the investment and operate the system. With an output of 10 h day-1 over a 10-year amortization period, the value of the energy output (R$183.69) was lower than the results obtained.

Brasil (2012) reports that renewable sources of electricity are likely to increase significantly in 5 years, because the cost of implementing such sources is decreasing. In the auction held by the federal government for the purchase of wind energy, the megawatt-hour was negotiated at R$180.00. However, solar energy is not at a competitive level; the price of a megawatt-hour is between R$300.00 and R$400.00. The high cost is due to the high initial investment to produce solar energy.

Chynoweth, (2004); Walla & Schneeberger, (2009) when analyzing the viability of biogas plants found a mixture of relevant variables: the economic efficiency of anaerobic digestion, among others such as investment cost, operating cost of the biogas plant and methane production, relevant factors that were considered in this study.

With the sale of carbon credits, the value of energy production is drastically reduced. With payback times of 10, 15 and 20 years, generating 10, 16 and 20 h.d-1, there is a cost reduction of approximately 60% with the sale of carbon credits. Producing 10h, 16h and 20 h, carbon credits earn R$ 43,238.38 - R$ 56,921.41 - R$ 71,151.76 respectively.

Table 6 shows the cost of energy production considering the investment in the biodigester at 30% of the total value, considering only the installation of the vinimanta, without the sale of carbon credits. In 15 years and 10h d^{-1} , the cost of producing electricity (R$ 108.64 MWh^{-1}) becomes viable for the producer. The other conditions shown in Table 2 all demonstrate the viability of producing electricity from pig biogas.

Table 6 - Cost of electricity (R$.MWh-1) from pig biogas, varying the production time as a function of the useful life of the biodigester, with an investment of 30% of the total value of the biodigester.

Biodigester lifespan (years)	Energy production time h.day^{-1} without selling carbon credits		
	10	1620 h	
10	R$ 143,45	R$ 97,43	R$ 87,34
15	R$ 108,64	R$ 73,74	R$ 66,13
20	R$ 91,92	R$ 62,42	R$ 55,96

With regard to the NPV, Figure 7 shows the values in relation to the payback time for the investment and operation of the engine generator, without the sale of carbon credits.

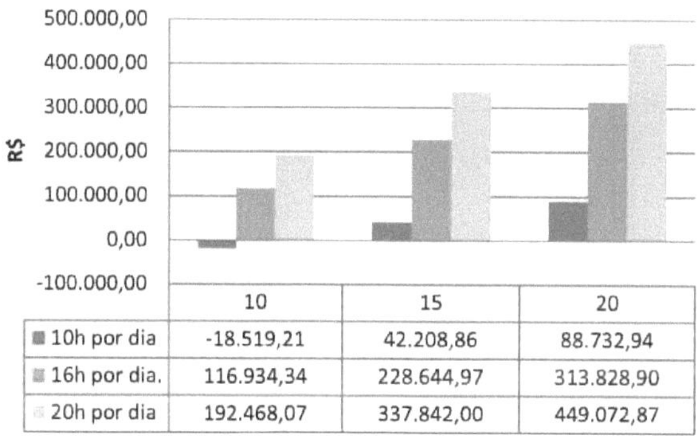

Figure 7-VPL as a function of TRI and motor-generator operation considering 30% of the biodigester's total value

In the situation where the NPV is negative (10 years / 10h d^{-1}), the implementation of the electricity generation system is not feasible at 5.5% p.a., which would result in a loss to the producer of R$ 18,519.21.

Figure 8 shows the negative NPV in various situations, considering the total value of the investment. The NPV becomes positive from the situation of 15 years of TRI operating 20h d^{-1} . However, the NPV becomes negative again when operating 10h d^{-1} with 20 years of TRI. From then on, the NPV becomes positive again. In other words, in most of the situations analyzed, the producer will make a loss if he produces energy from pig biogas. Gwavuya (2012), analyzing the collection of manure for the production of biogas and energy for households, found that the NPV was positive over a 20-year payback period, considering a discount rate of 4%.

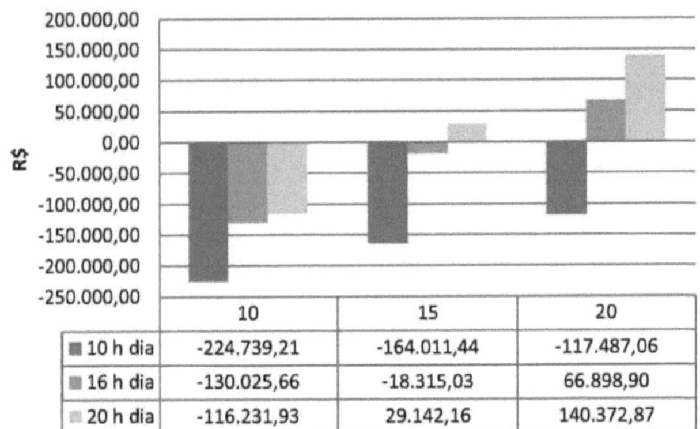

	10	15	20
10 h dia	-224.739,21	-164.011,44	-117.487,06
16 h dia	-130.025,66	-18.315,03	66.898,90
20 h dia	-116.231,93	29.142,16	140.372,87

Figure 8 - NPV as a function of TRI and motor-generator operation considering the total value of the biodigester without the sale of carbon credits.

Figure 9 shows the NPV as a function of payback time, considering the total value of the investment with the sale of carbon credits.

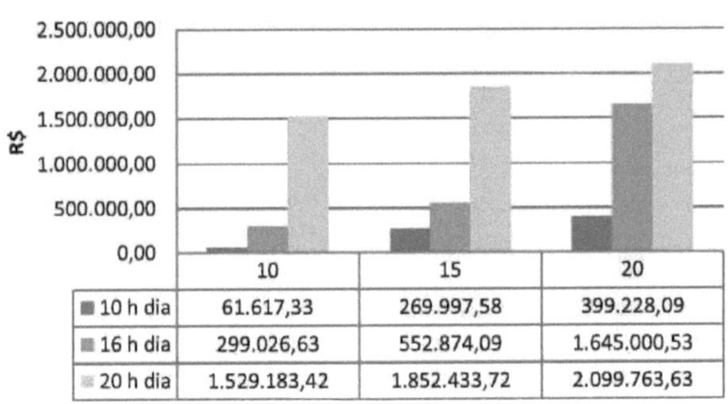

	10	15	20
10 h dia	61.617,33	269.997,58	399.228,09
16 h dia	299.026,63	552.874,09	1.645.000,53
20 h dia	1.529.183,42	1.852.433,72	2.099.763,63

Figure 9 - NPV as a function of TRI and motor-generator operation considering the total value of the biodigester with the sale of carbon credit

When there is income from the sale of carbon credits, all the situations analyzed generate a positive NPV, generating a profit for the producer. This is because the carbon market is an area that is expanding all over the planet, due to the fact that it contributes to the environment and at the same time generates income for small and large food producers.

It doesn't just cover animal husbandry, but all renewable energy sources in general.

The cost of producing biogas varies according to the amount of biogas produced and the initial value of the investment. The cost of biogas varies from R$ 0.25 to R$ 0.15.

Table 7- Cost of biogas production (R$ m-3)

	10 years	15 years	20 years

10 hours	R$ 0,25	R$ 0,19	R$ 0,16
16 hours	R$ 0,23	R$ 0,17	R$ 0,15
20 hours	R$ 0,23	R$ 0,17	R$ 0,15

DeVuyst et al. (2011) studied the difference between alternative energy production, comparing ethanol, biogas and cattle manure, varying the interest rate in two scenarios. The first with a rate of 4.5% and the second with a rate of 6%. In both, the calculated NPV was negative. The conclusion was that even though the NPV was negative, it is still feasible to install a biodigester to produce and sell surplus electricity, rather than just to treat agricultural waste.

Kunz et al. (2009), studying a pig manure treatment system, found that selling carbon credits reduces production costs by 14 to 18%. When a manure treatment system is installed on rural properties, it has the potential to generate additional income from biogas and biofertilizers. These revenues, however, represent a positive net present value, with an internal rate of return (IRR) ranging from 6.4% to 28.4% per year, depending on the environmental goals and the scale of the project.

Làzaro and Rud-strom (2007) evaluated the economics of an anaerobic digester for dairy farming. They concluded that electricity prices do not justify the installation and operation of a digester. Such results should vary significantly, however, according to the cost of any energy purchased, the price of energy sold and the state of regulations or utility agreements on metering. Stokes et al. (2008) evaluated the methane savings of electricity-generating biodigesters using real options for dairy cattle. They also concluded that the investment is not economically justifiable.

The profitability index is shown in Table 8. It can be seen that with an energy production of 10 h d^{-1} , there is no profit in any of the situations, confirming the results obtained with the NPV values. When producing 16 h d^{-1} , the profitability index becomes greater than 1, with a return on investment of 20 years. With a generation time of 20 h of^{1} , the index is less than 1 with a payback time of 10 years. In the other situations, the index is greater than 1, indicating a return on investment and profit for the producer.

Table 8 - Profitability index

10 h			16 h		20 h	
	S/ CC	C/CC	S/CC	C/CC	S/CC	C/CC
10 years	0,45	1,15	0,72	1,64	0,79	3,76
15 years	0,60	1,66	0,96	1,66	1,05	4,34
20 years	0,71	1,98	1,14	4,53	1,25	4,79

5. CONCLUSIONS

The production of electricity from pig biogas is not financially viable. It becomes economically viable when there is income from the sale of carbon credits.

In terms of payback time, it can be concluded that regardless of whether you generate 16 or 20 h^{-1} , the payback time is practically the same. This is because the initial investment to produce energy by generating 20 h of^{-1} is high. It is therefore recommended that producers only generate 16 h d^{-1} because the initial investment is lower.

The cost of producing electricity becomes competitive (without selling carbon credits) when the return on investment is over 15 years, generating energy for at least 16 $hours^{1}$. When carbon credits are sold, starting from a payback time of 10 years and generating 10 h d^{-1} , the cost of production is viable for the producer, generating income for the producer from the sale of surplus electricity.

Chapter III - Energy balance of biogas production in a pig finishing system

Summary

The technological innovation that has taken place in Brazilian agriculture in recent years and the consequent increase in the demand for energy inputs means that the issue of energy efficiency has become important in determining the degree of sustainability of food production. The production of electricity from alternative sources is increasingly gaining ground on the national and world stage, given that energy sources are predominantly fossil-based and finite. In this context, the energy balance of the production system for finishing pigs and its consequent production of electricity via biogas was analyzed. The largest share of input energy was obtained from feed consumption with 1,065,444 kg cycle^{-1} (79.13%). In terms of energy output, biofertilizer accounts for 44.70%, equivalent to 666,182.9 kg cycle^{-1} of all energy output. The energy coefficient was 0.52, making the system efficient in terms of energy production and consumption.

1. INTRODUCTION

Nowadays, the aim is to develop agriculture in a way that doesn't harm the environment and at the same time increases food production. The green economy requires innovations on a sustainable basis in order to be placed on competitive markets.

Greenhouse gas emissions, climate change and the instability of existing energy prices are important items to be studied in the search for a sustainable economy. Agriculture is one of the most important activities in this context. As well as producing food, it can, if poorly managed, pollute rivers and springs, cause forests to be cut down, and emit greenhouse gases (through animal production), among other aggravating factors.

Biogas production is one of the alternatives for reducing greenhouse gas emissions, and is being widely studied and used as pilot projects in several states with the encouragement of the country's private and public sector. As pig production has expanded in Brazil, so has the amount of waste generated, causing a huge environmental problem.

However, solutions have been implemented, such as stabilization ponds and biodigesters. They treat the waste, turning it into biogas or biofertilizer. The biogas can be used to generate electricity, and the biofertilizer can be used on crops. The energy balance of pig production becomes an important point for assessing environmental sustainability, pointing out where the greatest amount of non-renewable energy is spent, making it possible to find savings strategies.

The aim of this chapter is to calculate the energy balance of a pig farm that generates electricity and biofertilizer from biogas.

2. LITERATURE review

The development of agriculture since World War II has aimed to increase production through the use of inputs whose raw material is non-renewable resources, such as oil derivatives, with direct consequences for sustainability, not only in the economic field, but also in the energy flows involved (ALMEIDA et al., 2010).

Several countries in recent years have been using instruments such as economic incentives to increase the production of renewable energy. The objectives are to reduce greenhouse gas emissions and increase energy security by replacing fossil fuels. The political arguments for increasing the production of renewable energy are economic incentives and are classified into price and quantity measures (Shaw et al. 2010).

The energy used in agricultural systems, its distribution, flows and conversion make it important to assess the degree of sustainability of such systems, especially considering the crises in the Brazilian energy sector. This makes it possible to determine which processes, equipment and materials require the most energy, pointing to ways of saving energy (TEIXEIRA et al. 2005; CAMPOS et al. 2003). In order to evaluate family farming production systems, these analyses aim to provide a better understanding of the degree of sustainability and to determine their dependence on extra energy on the property, and the weight that this dependence has in the production process. (ALMEIDA et al. 2010).

The renewable economy is designed to develop bioenergy in order to reduce dependence on fossil fuels and consequently reduce greenhouse gas emissions and foster the development of the rural economy. To be considered a viable substitute for fossil fuels, an alternative fuel must exhibit environmental benefits superior to those of fossil fuels. It must be economically competitive and capable of being produced to meet energy demand. It must also present a net energy gain over the energy sources used to produce it (Bridgwater, 2006).

The energy balance aims to determine energy flows, point out their total demand, their energy efficiency shown by the net energy gain and the output/input ratio (energy produced/energy consumed) and the energy used to produce or process one kilogram of a given product (Siqueira et al. 1999; Pimentel, 1980). Considering the importance of sustainable food production, the energy balance and economic study become indicative of environmental sustainability, considering the use of non-renewable energies and sustainability in the field as a requirement for the continuation of agricultural activity (Pracucho et al. 2007).

Agricultural production with its intensive systems has caused serious environmental damage in two ways: one due to the increasing and agile depletion of natural resources and the other due to pollution or contamination caused by the excessive release of waste components into the environment (Romero et al. 2008; Kosioski and Ciocca, 2000). Every food production process generates a large amount of waste and this waste stores some energy. Some systems can convert this waste into energy, reducing its production cost and making it work in an energy-balanced way (Santos and Lucas Júnior, 2004).

The energy balance is directly related to the economic balance and its relevance has been studied periodically. Souza et al. (2012) evaluated the life cycle of sugarcane ethanol and found that the agricultural phase requires 96% of all energy for ethanol production, the largest contribution (48.5%) being associated with the use of fossil fuels for harvesting and transporting agricultural production. Poschl et al. (2010) studied the energy efficiency of biogas production in various ways and found that energy efficiency can be increased by up to 6.1% with residual biogas recovery from biodigesters.

The potential use of biodigesters for fertilizer production reduces dependence on energy-intensive mineral fertilizers, thereby reducing greenhouse gas emissions (ENVITEC, 2009; Tambone et al, 2009). The use of biogas results in a positive life cycle energy balance. Energy balance analyses of systems that use biogas as a raw material have been updated frequently, but there is no reliable database for comparisons due to the variability of systems (Berglund & Borjesson, 2006).

2.1 Pig production and anaerobic digestion

Every year, millions of tons of solid waste are produced from municipal, industrial and agricultural sources. The indiscriminate decomposition of this organic waste leads to contamination of the land, water and air. Of all the forms of organic solid waste, the most abundant is animal manure generated from small farms to large farms, and as the number of animals increases, the more intense the pollution problem becomes (Nasir et al. 2012).

The production of biogas through the anaerobic digestion of organic matter produces residues (biofertilizers). When applied to agricultural soil, it emits even small amounts of CO2 and CH4 (Gerardi, 2003). The number of biogas production units has increased significantly in the last decade, resulting in an increase in the amount of waste produced. Therefore, finding a sustainable, economical and safe way of using this waste is of the utmost importance (Odlare et al. 2012).

Angonese et al. (2006) and Pereira et al. (2008) point out that with the expansion of pig farming in the country and technological advances in production, the generation of waste has increased and much of this waste is discharged into rivers and springs. Due to the adoption of confined pig production systems, a large amount of waste is generated. According to Souza et al (2004), one pig can produce 7.2 liters of waste a day, considering that the efficiency of the process is around 60.5%, the production of methane, taking into account the organic load, is approximately 0.504 m^3 .head d^{-1-1} , which is equivalent to a biogas production of 0.775 m^3 of biogas.head of pig^{-1} .d $.^{-1}$

The production of biogas through biomass digestion produces carbon dioxide as a product. Together with the hydrogen produced by electrolysis, this could provide a renewable source of energy. The catalytic reduction of carbon dioxide could produce biogas by using the carbon in the biomass feedstock efficiently (Mohseni et al. 2012).

As a product of anaerobic digestion, biogas is a potential candidate to replace fossil fuels and reduce methane emissions into the atmosphere. As a substitute for diesel in motor-generators, it has been generating quality electricity to meet the consumption of both small and large properties. Another product of anaerobic digestion is biofertilizer, which is used on crops, reducing production costs compared to conventional fertilizers.

The treatment of organic waste through anaerobic digestion processes has been recognized as a means of controlling the greenhouse effect and generating electricity (Barton et al. 2008). In this respect, several countries have agreed to subsidize the production of biogas as a renewable energy source in the combined production of heat and power, in order to reduce greenhouse gas emissions in line with the Kyoto Protocol (CCE, 2001).

In anaerobic digestion, around 50 to 90% of the organic material is converted into biogas which is removed from the reactor and a small part of the organic material (5 to 15%) is transformed into microbial biomass (Arruda, 2004). There is a wide range of waste that can be treated anaerobically (both rural, urban and industrial waste) with the aim of removing the polluting organic load and pathogenic microorganisms, producing stable biogas and biofertilizers, richer in nutrients and with better sanitary quality compared to the original material (Mittal, 2006; Merzouki et al., 2005; Parawira, et al., 2006).

According to Figure 10, the agroenergy matrix can be arranged as follows:

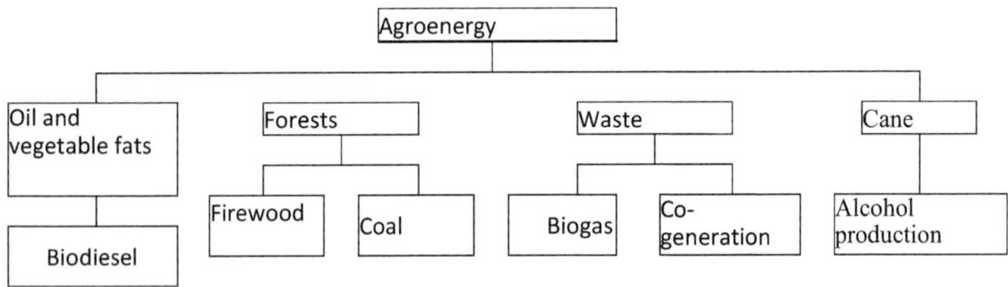

Figure 10 - Agroenergy Matrix. SOURCE: Ministry of Agriculture, Livestock and Supply (2006).

Agro-energy encompasses various sectors of food production, such as meat production, vegetable production (soy, corn), the food agro-industry and others. It has become important for the variation in the national energy matrix. As sustainability in food production is one of the most important issues today, the energy balance becomes essential for assessing the viability of food production.

The development of the renewable energy market differs from region to region. One example is Africa, which has very low rates of access to renewable energy, while in Latin America electricity comes basically from renewable sources. Rural renewable energy is a highly dynamic and constantly evolving market, but it is challenged by its lack of structure. There have been few public policy incentives for many years, but the implementation of new renewable energy production technology and cost reductions point to a prosperous future in the area.

Food production is tending to decrease relatively quickly, while the world's population is tending to increase, so we need to find ways of producing food that doesn't harm the environment and also contributes to the country's energy production, making it viable and environmentally friendly.

3. MATERIALS AND METHODS

3.1 System delimitation

The system was delimited by the activities related to pig production. The energy inputs are water, feed and electricity. The live weight of the piglets produced, biofertilizer and biogas are output energies, as specified in Figure 11. The final production of electricity was not taken into account because it is directly related to the production of biogas.

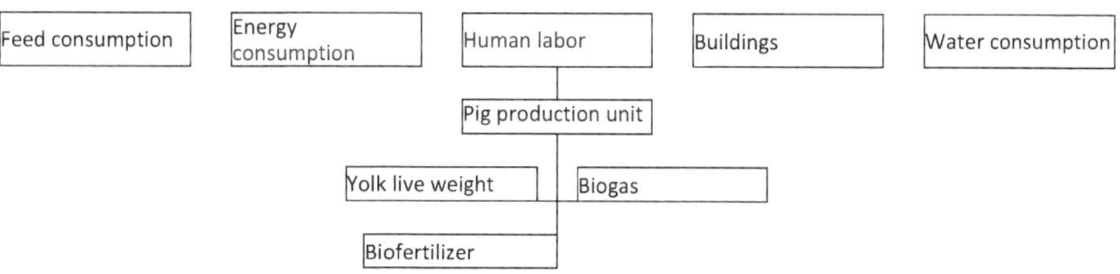

Figure 11 -- Colombari Farm's pig production system

3.2 Pig production system

The pig production system is intensive, i.e. a confined breeding system in the finishing phase. It has an average capacity of 4673 head. The animals are arranged in 42m stalls^2 , with 42 animals each. The animals arrive weighing 25 kg and finish the 120-day cycle weighing 120 kg .$^{-1}$

3.3 Waste treatment system

The manure treatment system consists of two Canadian model biodigesters in series. The first has a capacity of 29.2 m d$^{3-1}$, with a hydraulic retention time of 30 days. The second biodigester has a capacity of 7.3 m^3 d^{-1} , designed with a hydraulic retention time of 30 days, because the average temperatures in the coldest period of the year are between 16 and 17°C. At the outlet pipe, the effluent from the biodigester is disposed of in a manhole and taken to the manure house to store the biofertilizer, which is then distributed to the crops.

1.4 Energy matrix

The study focuses on the production of pigs in the finishing phase. The average feed consumption per animal is 1.9 kg of^{-1} animal^{-1} in the finishing phase. The feed components are crushed corn, soybean meal and nutritional core.

Water consumption is 1.8 L d^{-1} animal^{-1} . To calculate electricity consumption, an average of the last four months was taken.

To calculate the production of biofertilizer, the density presented by Sediyama et al. (2009) was used, which characterized pig biofertilizer as having the following properties after a retention time of 30 days: : N = 22.7; P = 15.2; K = 11.0; Ca = 17.0; Mg = 7.7; S = 3.9; C. org. = 2.1 and Na = 5.2 and, in mg L-1; , Zn = 2068; Fe = 3859; Mn = 176; Cu = 1166; Cr = 0.13; Ni = 0.21 and Cd = 0.01; pH (H_2O) = 8.61; density = 1.1 g cm3 and C/N = 0.09.

For human labor, it is assumed that six men work 8 hours a day for 120 days. The energy coefficients are shown in Table 9. Table 9 - Energy coefficients for input and output items

Entry		Exit	
Feed[(1)]	17 MJ.kg^{-1}	Piglets[w]	9.21 MJ.kg^{-1}
Water (2)	2.47 x 10^{-3} MJ.l^{-1}	Biofertilizers[(5)]	8 MJ.kg^{-1}
Energy consumption	13.11 MJ.kWh^{-1}	Biogàs(6)	22.35 MJ. m^{-3}
Electrical[(3)]			
Human work[(7)]	4.39 MJ h^{-1}		
Buildings	956.03 MJ m^{-2}		

Source: 1 OETTING (2002); 2 SANTOS & LUCAS JUNIOR (2004); 3 Brasil (2007); 4 COMITRE (1995); 5 PELLIZZI (1992); 6 ROPPA (2000); 7 PIMENTEL (1980).

1.5 Electricity production

The electricity generation system consists of a motor-generator set, protection system and command control, which is interconnected to the power distribution network. It has a 75 kVA (220/127 V) transformer, with a primary voltage of 13.8 kV and a 220/127 V output of 5 surrounding columns, with a 200 ampere circuit breaker.

The motor generator produces 80 kWh to meet the property's energy needs, such as the motor-pump set and the distribution of energy to the fertigation plant, the feed mill and the four homes on the property. The motor generator works according to the availability of biogas in the biodigesters, and has been sized to operate 8 hours a day.

1.6 Energy efficiency coefficient

The energy efficiency coefficient (η) is calculated using the methodology of Quesada et al. (1991) cited by Angonese et al. (2006) where:

$$\eta = \sum \frac{energia_{saída}}{energia_{entrada}} \tag{14}$$

Efficiency is measured by the energy balance or input/output ratio, which is achieved by determining the amount of energy obtained in the product in relation to that used in the system to produce it (Campos, 2004). The energy output is obtained by directly converting the product yield into energy (Albuquerque et al. 2008).

4. RESULTS AND DISCUSSION

Table 10 shows the amount of inputs needed for pig production. The largest share is in feed consumption. This is followed by water and electricity.

Table 10 - Quantity of inputs needed for pig production

Entry	Total
Feed consumption	1,065,444 kg $cycle^{-1}$
Water consumption	1,009,368 L $cycle^{-1}$
Electricity consumption	271,592 kWh $cycle^{-1}$
Human work	5760 h $cycle^{-1}$
Buildings	4673 m^2 $cycle^{-1}$

With regard to production at the end of the cycle, it can be seen that 560,760 kg^{-1} of pork, 666,182 kg^{-1} of biofertilizer and 55,360 m $^{3-1}$ of biogas are produced, as shown in Table 11.

Table 11 - Quantity of energy produced at the end of the cycle

Exit	Total
Beds produced	560,760 kg $cycle^{-1}$
Biofertilizers	666,182.9 kg $cycle^{-1}$
Biogàs	66,360 m^3 $cycle^{-1}$

The energy embedded in the input processes is shown in Figure 12. The largest proportion of energy in the input comes from feed consumption, which accounts for 79.13% of all energy in the input. Water consumption and human labor have the least embedded energy (0.010% and 0.049% respectively). With regard to energy consumption and buildings, the embedded energy is 1.29% and

19.51% respectively.

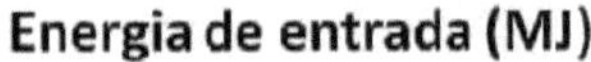

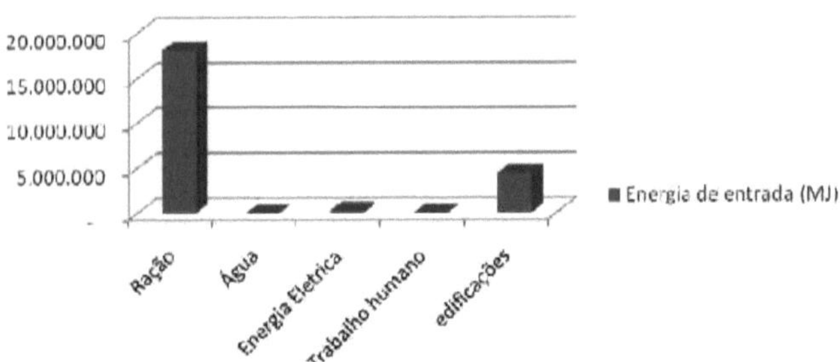

Figure 12 - Energy efficiency of the input components for the production cycle in the pig finishing phase

However, the largest proportion of energy output is linked to biofertilizer (44.70%), followed by pig production (42.85%) and biogas (12.44%), as shown in Figure 13.

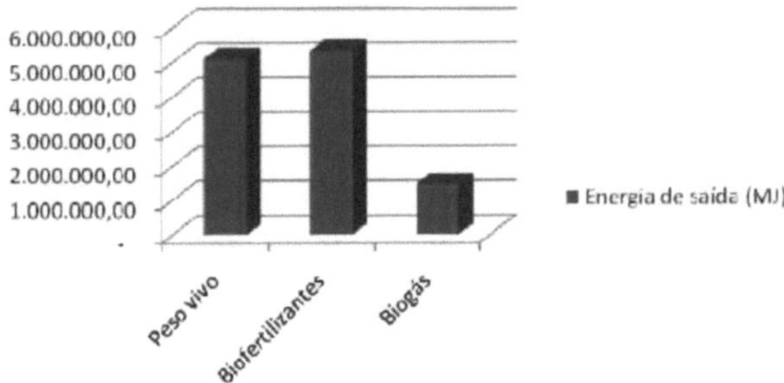

Figure 13 - Energy efficiency of the output components for the production cycle in the pig finishing phase

Of all the energy used in the production system, input energy accounts for 65.75% (22,890,573 MJ) of the total. The energy output is 34.24% (11,921,132 MJ). The energy efficiency coefficient found was 0.5208. Souza et al. (2009) analyzed the pig production system together with pasture development and found an energy coefficient of 0.31. Angonese et al. (2006) analyzed a pig finishing cycle and obtained a coefficient of 0.38.

According to Poschl et al. (2010), the variation in the energy coefficient depends on the system being analyzed and the quality of the raw material.

Farrell et al. (2006) and Hammerschalag (2006), studying the energy balance of ethanol production using corn as a raw material, concluded that the return on energy invested is positive and its use significantly reduces the use of oil. Reducing fossil fuel inputs is still advantageous, especially in terms of reducing greenhouse gas emissions. To improve the energy balance, it is necessary to reduce the input of fossil fuels and use the flows of various by-products in animal feed (DeVuyst et al. 2011).

5. CONCLUSIONS

Breed consumption is the one that demands the greatest amount of energy at the system's input. This corresponds to 79.13%. This is due to the composition of the feed and the amount consumed per animal.

Biofertilizer accounts for 44.70% of the energy output, followed by live weight and biogas.

The energy coefficient of the system studied is 0.5208, i.e. 52.08% of the input energy is converted into useful energy at the end of the cycle.

6. GENERAL CONCLUSIONS

With the need to renew the energy matrix, governments are encouraging new forms of energy production with a view to economic development and sustainable food production.

Producing electricity on the property is not economically viable from a financial point of view. It only becomes viable when the producer earns income from selling carbon credits.

Regardless of whether you generate 16 hours of^{-1} or 20 hours of^{-1} , the payback time is practically the same. This is due to the high initial investment for generating 20 h^{-1} . Therefore, it is recommended that producers only produce 16 h^{d-1} , because the initial investment is lower.

The cost of producing electricity becomes competitive when the payback time is over 15 years with a minimum generation of 16 h d^{-1} . With the sale of carbon credits, the payback time drops to 10 years and with a generation of 10 h d^{-1} , energy production becomes viable.

The energy balance shows energy efficiency in the pig production system. The largest portion of energy input corresponds to feed consumption. Biofertilizer accounts for the largest share of energy output.

The Colombari farm is considered a benchmark for sustainability in the western region of Paranâ. Its pilot project is yielding good results in terms of sustainability in food production. However, when a deeper analysis of the system is made, there are indications that the production system is not viable. But it is an example for other pig, cattle and poultry producers to follow, as it is a clean and renewable energy production system.

7. BIBLIOGRAPHICAL REFERENCES

ABIPECS - **Brazilian Association of the Pork Producing and Exporting Industry** (2012).

ALBUQUERQUE, F. A. et al. **Energy analysis of the castor bean and peanut consortium**. In:

CONGRESSO BRASILEIRO DE MAMONA, 6., 2008, Salvador. Proceedings... Salvador: SEAGRI: Embrapa Algodao, 2008a. 1 CD-ROM.

ALMEIDA L.C. F de., BUENO O. de C., ESPERANCINI M.S.T. **Energetic evaluation of the corn crop in a rural settlement**. Iperó-SP. Revista Energia na Agricultura, Botucatu, v.25, n.2, p.105-122, 2010.

ANEEL - National Electric Energy Agency. **Atlas of Electricity in Brazil**. 2008.

ANGONESE A. R., CAMPOS A. T., ZACARIM C. E., MATSUO M. S., CUNHA F. **Energy efficiency of a pig production system with waste treatment in biodigesters**. Revista Brasileira de Engenharia Agricola e Ambiental, Campina Grande, v.10, n.3, p.745-750, 2006.

ANGONESE A. R., CAMPOS A. T., WELTE R. A. **Potential for reducing carbon equivalent emissions from a pig farm using a biodigester**. Engenharia Agricola, Jaboticabal, v.27, n.3, p.648-657, 2007.

ARRUDA, V. C. M., **Anaerobic treatment of effluents generated in cattle slaughterhouses**. 2004. 127 f. Dissertation (Master's Degree in Civil Engineering) Federal University of Pernambuco - UFPE, Recife-PE.

ARTHURSON V. **closing the global energy and nutrient cycles through application of biogas residue to agricultural landpotential benefits and drawbacks.** Energies, v.2, p.226-40, 2009.

BEN - **National Energy Balance 2011: base year 2010** / Empresa de Pesquisa Energética. - Rio de Janeiro : EPE, 2011.

BERGLUND M, BORJESSON P. **Assessment of energy performance in the lifecycle ofbiogas production**. Biomass Bioenergy; v.30, p.254-66, 2006.

BIOGASBURNER. **Design equations for gas burner**. 2012

BORJESSON, P., BERGLUND, M., 2003. **Environmental analysis of biogassystems.** Report nr. 45, Department of Environmental and EnergySystems Studies, Lund University, Lund, Sweden, (Report, in Swedishwith English summary).

BORJESSON P., BERGLUND M**. Environmental systems analysis of biogas systems - Part i: Fuel-cycle emissions**. Biomass and Bioenergy, v.30, p.469-485, 2006

BRAZIL - Ministry of Agriculture, Livestock and Supply. **Low-carbon** agriculture**.** 2012.

BRAZIL. Ministry of Mines and Energy. **National Energy Balance**. Brasilia, 2007. 192p.

BRAZIL - **Portal Brasil**. 2012.

CAMPOS, A.T.; SAGLIETTI, J.R.C.; CAMPOS, A.T.; BUENO, O.C.; RESENDE, H.;

GASPARINO, E.; KLOSOWSKI, E.S. **Energy cost of building a hay storage facility.** Ciência Rural, Santa Maria, v.33, n.4, p.667672, 2003.

CAMPOS, A. T. **Agricultural energy balances: an important tool to indicate the sustainability of agroecosystems**. Revista do Centro de Ciência Rural, v. 34, n. 06, p. 1977-1985, 2004.

CCE - Energy Conservation Center. **Biogàs Technical Guide**. Ed. JE92 Projetos de Marketing Ltda., Alges, June 2000.

CCEE - Electricity Trading Chamber. **3rd Reserves Auction and 2nd Renewable Sources Auction.** 2010.

CHEN S., CHEN B., SONG D. **Life-cycle energy production and emissions mitigation by comprehensive biogas-digestate utilization**. Bioresource Technology, v.114, p.357-364, 2012.

CHYNOWETH, D.P. (2004). **Biomethane from energy crops and organic wastes**. International Water Association, Anaerobic Digestion 2004, Proceedings 10th World Congress, vol. 1, Montreal, Canada, pp. 525-530.

COLDEBELLA A. **Feasibility of Electricity Cogeneration with Dairy Cattle Biogas.** Dissertation (Master's Degree in Agricultural Engineering). State University of Western Paranà. 2006.

COLLET P., HÉLIAS A., LARDON L., RAS., M., , GOY R. A STEYER., J. P. **Lifecycle assessment of microalgae culture coupled to biogas production.** Bioresource Technology, v.102, p.207-214, 2011.

COMITRE V. **The energy issue and the technological pattern of Brazilian agriculture. Informaçoes econômicas**, Sao Paulo, v.25, n.12, p.29-35, 1995.

COPEL - **Distributed Generation Program with Environmental Sanitation**. 2009.

CORNELISSEN ST., KOPER M., DENG Y. Y. **The role of bioenergy in a fully sustainable global energy system**. Biomass and Bioenergy, v.41, p.21-33, 2012.

CORTEZ, L.A.B., LORA, E. E. S., GOMEZ, E. **Biomass for energy**. Campinas: UNICAMP, 2008

COSTA D. F. da. **Generating electricity from sewage treatment biogas**. Sao Paulo, University of Sao Paulo, 2006. Dissertation. 194p.

DAM J. V, JUNGINGER M., FAAIJ A., JÜRGENS I., BEST G., FRITSCHE U. **Overview of recent developments in sustainable biomass certification**. Biomass and Bioenergy, v.32, p.749-780, 2008.

DEUBLEIN, D; STEINHAUSER, A. **Biogas from waste and renewable resources: an introduction**. Weinhein-Germany: Verlag GmbH & Co. KGaA, 2008.

DEVUYST E. A., PRYOR S. W. , LARDY G. EIDE W., WIEDERHOLT R. **Cattle, ethanol, and**

biogas: Does closing the loop make economic sense? Agricultural Systems, v.104, p.609-614, 2011.

DURÀES, F. O. M. **Biofuels: real issues for Brazil's sustainable development equation**. Revista de Politica Agricola, n. 1, p. 129-134, 2008.

ERB K. H., HABERL H., PLUTZAR C. **Dependency of global primary bioenergy crop potentials in 2050 on food systems, yields, biodiversity conservation and political stability**. Energy Policy, v.47, p.260-269, 2012.

FAO. **Opportunities and challenges of biofuel production for food security and the environment in Latin America and the Caribbean**. Document prepared for the 30ª Session of the FAO Regional Conference for Latin America and the Caribbean, Brasilia, Brazil; April 14-18, 2008.

FARRELL, A.E., PLEVIN, R.J., TURNER, B.T., JONES, A.D., O'HARE, M., KAMMEN, D.M., **Ethanol can contribute to energy and environmental goals**. Science, v. 311, p.506-508, 2006

FERNANDES D. M., **Biomass and Biogas from Pig Farming**. 2012. 176p. Dissertation (Master's Degree in Energy in Agriculture). State University of Western Paranà. Cascavel 2012.

GBEP. **A review of the current state of bioenergy development in G8 + 5 countries.**

Global Bioenergy Partnership; 2007

GERARDI M.K. **The microbiology of anaerobic digesters**. Hoboken NJ, U.S.A: John Wiley & Sons, Inc; 2003. p. 177. .

GOES, T.; ARAÙJO, M.; MARRA, R. Biodiesel **and its Sustainability. 2010.** (EMBRAPA technical articles) .

GULOSIN, M., OSTOJIC, A., LATINOVIC S., JANDRIC M., IVANOVIC O.M., **Review of the economic viability of investing and exploiting biogas electricity plant - case study** Vizelj, Serbia. Renewable and Sustainable Energy Reviews, Volume 16, Issue 2, February 2012, pages 1127-1134.

GWAVUYA S.G., ABELE S., BARFUSS I., ZELLER M., MÜLLER J. **Household energy economics in rural Ethiopia: A cost-benefit analysis of biogas energy**. Renewable Energy, Volume 48, December 2012, Pages 202-20.

GOLDEMBERG, J., LUCON, O., **Energy, Environment and Development**. 3 ed. Sao Paulo: University of Sao Paulo, 2008.

HAMMERSCHLAG, R., **Ethanol's energy return on investment: a survey of the literature 1990 - present**. Environmental Science and Technology, v.40, p.17441750, 2006.

HONÓRIO M. O. **Estimation of carbon credit from the production and burning of biogas from swine manure: Case study**. 88p. Dissertation (Master's Degree in Chemical Engineering). Federal

University of Santa Catarina, Florianópolis, 2009.

IANNICELLI, L. A. **Energy reuse of biogas from a brewing industry**. Taubaté. Master's dissertation. DEM/ UNITAU, 2008

IBGE - **Brazilian Institute of Geography and Statistics** - AEB. Rio de Janeiro: IBGE, 2006.

ICB - **Instituto Carbono Brasil**. 2012.

JUNGINGER M., BOLKESJ T., BRADLEY D., DOLZAN P., FAAIJ A., HEINIMO J., HEKTOR B., LEISTAD O., LING E., PERRY M., PIACENTE E., ROSILLO-CALLE F., RYCKMANS Y., SCHOUWENBERG P. P., SOLBERG B., TROMBORG E., WALTER A. S., WIT M. DE. **Developments in international bioenergy trade.** Biomass and Bioenergy, v.32, p.717-729, 2008.

KARPENSTEIN-MACHAN M. **Sustainable cultivation concepts for domestic energy production from biomass**. Critical Reviews in Plant Sciences, v.20, p.1-14, 2001.

KEY, N., STACY **Sneeringer. Climate Change Policy and the Adoption of Methane Digesters on Livestock Operations**, ERR-111,U.S. Department of Agriculture, Economic Research Service, February 2011.

KUNZ A., MIELE M., STEINMETZ R.L.R. **Advanced swine manure treatment and utilization in Brazil.** Bioresource Technology, v.100, p. 5485-5489, 2009.

KOSIOSKI, G.V.; CIOCCA, M.L.S. **Energy and sustainability in agroecosystems**. Ciência Rural, Santa Maria, v.30, n.4, p.737-745, 2000.

LANTZ, M., 2004. **Farm based biogas production for combined heat and power production: economy and technology**. Environmental and Energy Systems Studies, Lund University, Lund (Master's thesis, in Swedish with English summary).

LANTZ M., SVENSSON M., BORJESSON L., BORJESSON P. **The prospects for an expansion of biogas systems in Sweden-Incentives, barriers and potentials**. Energy Policy, v.35, p.1830-184, 2007.

LIEBRAND, C., LING K. C.. **"Carbon Credits for Farmers." US Department of Agriculture**. Rural Development. Rural Cooperatives, v.75, p.10-12, 2008

MALKKI H, VIRTANEN Y. **Selected emissions and efficiencies of energy systems based on logging and sawmill residues**. Biomass Bioenergy, p.321-327, 2003.

MARQUES C. A. **Micro-generation of electricity on a rural property using biogas as the primary source of electricity**. Cascavel, Unioeste, 2012. Dissertation. 81 p.

MARTHA JR., G. B. **Land use dynamics in response to sugarcane expansion in the Cerrado**.

Revista de Politica Agricola, n. 3, p. 31-44, jul./ago./set, 2008.

MARTINS, F. M. OLIVEIRA, P. A. V. de. **Economic analysis of electricity generation from biogas in pig farming.** Engenharia Agricola. v.31, p. 477-486, 2011.

MCTI. MINISTRY OF SCIENCE AND TECHNOLOGY AND INNOVATION. **Current status of clean development mechanism (CDM) project activities in Brazil and worldwide**. 2007.

MERZOUKI, M.; BERNET, N.; DELGENÈS, J.P.; BENLEMLIH, M. **Effect of prefermentation on denitrifying phosphorous removal in slaughterhouse wastewater.** Bioresource Technology, v.96, n.12, p.1317-1322, 2005.

MINISTRY OF AGRICULTURE, LIVESTOCK AND SUPPLY. **National Agroenergy Plan, 2006-2011**. 2ª Revised Edition. Brasilia, 2006.

MITTAL, G.S. **Treatment of wastewaters from abattoirs before land application - a review**. Bioresource Technology, v.97, n.9, p.1119-1135, 2006.

MME - Ministry of Mines and Energy. **National Energy Balance. 2011**.

MME - Ministry of Mines and Energy. PROINFA: **Source Incentive Program**

Energy Alternatives. 2005

MOLLER, H.B., NIELSEN, A.M., NAKAKUBO R., OLSEN, H. J. **Process performance of biogas digesters incorporating** pre-separated **manure. Livestock** Science v.112, p.217-223, 2007.

MOHSENI F., MAGNUSSON M., GORLING M., ALVFORS P.. **Biogas from renewable electricity - Increasing a climate neutral fuel supply**. Applied Energy, v.90, p.11-16, 2012.

NASIR I. M., GHAZI T. I. M., OMAR R. **Production of biogas from solid organic wastes through anaerobic digestion: a review**. Applied Microbiology and Biotechnology, v.95, p.321-329, 2012.

OETTING L. L., **Evaluation of different markers for determining digestibility and feed passage rate in pigs**. 2002. 59f. Dissertation (Master's Degree in Nuclear Energy in Agriculture) - Center for Nuclear Energy in Agriculture, University of Sao Paulo, Piracicaba, 2002.

ODLARE M., ABUBAKER J., LINDMARK J., PELL M., THORIN E., NEHRENHEIMA E. **Emissions of N2O and CH4 from agricultural soils amended with two types of biogas residues**. Biomass and Bioenergy, v.44, p.112-116, 2012.

OLIVEIRA, P.A.V. de; HIGARASHI, M.M. **Generation and use of biogas in pig production units**. Concórdia: Embrapa Swine and Poultry, 2006.

OLIVEIRA, P.A.V. de; MARTINS, F.M. **Utilization of biogas in pig farming to generate electricity**. In: BRAZILIAN CONGRESS OF AGRICULTURAL ENGINEERING, 36, 2007,

Bonito. Proceedings... Jaboticabal: Brazilian Association of Agricultural Engineering, 2007. 1 CD-ROM.

PARAWIRA, W.; MURTO, M.; ZVAUYA, R.; MATTIASSON, B. **Comparative performance of a UASB reactor and an anaerobic packed-bed reactor when treating potato waste leachate**. Renewable energy, v.31, n.6, p.893-903, 2006.

PEREIRA B. D., MAIA J.C.S., CAMILOT R. **Technical efficiency in pig farming: effect of environmental expenditures and tax exemption**. Revista Brasileira de Engenharia Agricola e Ambiental, Campina Grande, v.12, n.2, p.200-204, 2008.

PIMENTEL, D. **Handbook of energy utilization in agriculture**. Boca Raton: CRC Press, 1980. 475p.

POSCHL M., WARD S., OWENDE PHILIP. **Evaluation of energy efficiency of various biogas production and utilization pathways**. Applied Energy, v.87, p. 33053321, 2010.

PRACUCHO T. T. G. M., ESPERANCINE M. S.T., BUENO O.C. **Análisis energètica econômica da produçà de milho (Zea mays) em sistema de plantio direto em propriedades familiares no municipio de Pratânia - SP**. Energia na agricultura, Botucatu, v.22, p. 94-109, 2007.

PUKASIEWICZ S. R. M. **Treatment of effluent from the processing of by-products of the meat products industry in an anaerobic filter**. 69 f. (Master's Degree in Food Science and Technology). State University of Ponta Grossa, Ponta Grossa, 2010

REHL, T., MÜLLER, J. **Life cycle assessment of biogas digestate processing technologies.** Resources, Conservation and Recycling, v.56, p.92-104, 2011.

ROMERO, M.G.C.; BUENO, O.C.; ESPERANCINI, M.S.T. **Energy and economic efficiency in family cotton production systems.** Informaçoes Econômicas, Sao Paulo, v.38, n.1, p.7-19, 2008.

ROPPA L. **Update on Cholesterol, Fat and Calorie Levels in Pork,"** School of Veterinary Medicine, USP, 2000.

SANTOS, T.M.B.; LUCAS JÛNIOR, J. **Energy balance in a broiler house**. Engenharia Agricola, Jaboticabal, v.24, n.1, p.25-36, 2004.

SOUZA, S. N. M., **Production of electricity from biogas**. Institute of Applied Technology.2011.

SVENSSON, M., 2005. **The technology and economy of farm-scale, high solids anaerobic digestion of plant biomass**. Doctoral Dissertation, Department of Biotechnology, Lund University, Lund.

SHAW D., HUNG M. F., LIN Y. H **Using net energy output as the base to develop renewable**

energy. Energy Policy, v. 38, p.7504-7507, 2010.

SILVA C. R. A. DA., GARRAFA M. T. F., NAVARENHO P. L., GADO R., YOSHIMA S. **A biomassa como alternativa energética para o Brasil** (2005).

SILVA, E.P.; CAMARGO, J.C.; SORDI A.; SANTOS, A.M.R. **Recursos energéticos, meio ambiente e desenvolvimento**. Multiciencia, UNICAMP, v.I, 2003.

SINGH K.J., SOOCH S. S., **Comparative study of economics of different models of family size biogas plants for state of Punjab, India** - Energy Conversion and Management, Volume 45, Issues 9-10, June 2004, Pages 1329-1341.

SIQUEIRA, R.; GAMERO, C.A.; BOLLER, W. **Energy balance in the implantation and management of ground cover plants**. Engenharia Agricola, Jaboticabal, v.19, n.1, p.80-89, 1999.

SOUZA, S. N. M. de, PEREIRA, W. C. PAVAN, A. A. **Cost of electricity generated in a motor-generator set using pig biogas**. Acta Scientiarum Technology, V. 26, n.2, 2004. Pages 127-133.

SOUZA C. V., CAMPOS A. T., BUENO O. C., SILVA E. B. **Análisis energètica em sistema de produçâo de suinos com apro vei tamento de dejetos como biofertilizantes em pastagem**. Engenharia Agricola, Jaboticabal, v.29, n.4, p.547-557, 2009.

SOUZA S. P., AVILA M.T., PACCA S. **Life cycle assessment of sugarcane ethanol and palm oil biodiesel joint production**. Biomass e Bioenergy, v.44, p.70-79, 2012.

STARR K., GABARREL X., VILLALBA G., TALENS L., LOMBARDI L., **Life Cycle Assessment of Biogâs Upgrading Technologies**. Waste Management, v. 32, p.991-999, 2012.

STOKES, J.R., RAJAGOPALAN, R.M., STEFANOU, S.E., **Investment in a methane digester: an application of capital budgeting and real options. Review of Agricultural** Economics, v.30, p.664-676, 2008.

TAMBONE F, GENEVINI P, D'IMPORZANO G, ADANI F. **Assessing amendment properties of digestate by studying the organic matter composition and the degree of biological stability during the anaerobic digestion of the organic fraction of MSW.** Bioresource Technology v.100: p.3140-2. 2009;

TEIXEIRA, C.A.; LACERDA FILHO, A.F.; PEREIRA, S.; SOUZA, L.H.; RUSSO, J.R. **Energy balance of a tomato crop**. Revista Brasileira de Engenharia Agricola e Ambiental, Campina Grande, v.9, n.3, p.429-432, 2005.

VIOLA, E. **Brazil in the International Climate Change Mitigation Arena**. Center for Integration and Development Studies, Brasilia, 2009.

WARD A, HOBBS P.J., HOLLIMAN P.J., JONES D.L. **Optimization of the anaerobic digestion of agricultural resources**. Bioresource Technology, v.99, p. 7928-7968, 2008.

WALLA, C., SCHNEEBERGER, W. **The optimal size for biogas plants**. Biomass and Bioenergy v. 32, p.551-557, 2009.

WALSH, J.L., ROSS, C. C., SMITH, M.S., HARPER, S.R., WILKINS, W.A. **Handbook on biogas utilization. Georgia, Atlanta, USA**: Georgia Tech Research Institute (GTRI) and U.S. Department of Energy (DOE), 156p. 1988.

ZAGO, S. **Potential energy production through biogas integrated with environmental improvement in rural properties with intensive animal husbandry in the midwest region of Santa Catarina.** 2003. 103 f. Dissertation (Master's Degree in Environmental Engineering) - Universidade Regional de Blumenau, Centro de Ciências Tecnológicas, Blumenau, 2003.

8. ANNEXES

10 h production time^{-1} - lifetime	**- 10 /15/20 years**
Average annual biogas production	201.845 m^3
Average annual electricity production	241,703 kWh
Average annual electricity consumption	67898 kWh
Amount paid by the energy company	R$ 140.00 MWh
Interest rate	5.5% p.a.
Biodigester initial investment	R$ 294.600,00
Initial motor-generator investment	R$ 113.550,00
Methane density	0.67 kg ^{m3}
Percentage of methane in biogas	60%

16 h production time^{-1} - lifetime	**- 10 /15/20 years**
Average annual biogas production	265.720 m3
Average annual electricity production	386,608 kWh
Average annual electricity consumption	67898 kWh
Amount paid by the energy company	R$ 140.00 MWh
Interest rate	5.5% p.a.

Biodigester initial investment	R$ 352.800,00
Initial motor-generator investment	R$ 113.550,00
Methane density	0.67 kg ^{m3}
Percentage of methane in biogas	60%
Production of 20 hours^{-1} - lifetime	**- 10 /15/20 years**
Average annual biogas production	332.150 m3
Average annual electricity production	483,260 kWh
Average annual electricity consumption	67898 kWh
Amount paid by the energy company	R$ 140.00 MWh
Interest rate	5.5% p.a.
Biodigester initial investment	R$ 441.000,00
Initial motor-generator investment	R$ 113.550,00
Methane density	0.67 kg ^{m3}
Percentage of methane in biogas	60%

Printed by Books on Demand GmbH, Norderstedt / Germany